[redacted]

Magnetotellurics in oil exploration in the U.S.S.R.

I. M. Alperovitch
V. P. Busnov
D. A. Varlamov
Yu. K. Konnov
Yu. F. Konovalov
G. V. Koschlakov
G. A. Chernyaskii
A. L. Skeinkman
I. A. Yakovlev

VNII Geofisika, IGIRGI

Editors

Keeva Vozoff
Michael Asten
Macquarie University
North Ryde, N.S.W., Australia

English translation

George V. Keller
Colorado School of Mines
Golden, Colorado

SOCIETY OF EXPLORATION GEOPHYSICISTS

[redacted]

ISBN 0-931830-19-2
Library of Congress Catalog Number: 81-51441
Society of Exploration Geophysicists
P.O. Box 3098
Tulsa, OK 74101

Published 1982
Printed in the United States of America

FOREWORD

Our aim was to make the material available to the Western audience, rather than to rewrite the book. The work represents a decade of experience which differs from that in the West in terms of geologic environments, surface conditions, and of course technology. But, in common with us, the authors have learned to use this peculiar method (and it *is* peculiar) to advantage in the important business of petroleum exploration.

Keeva Vozoff, Editor

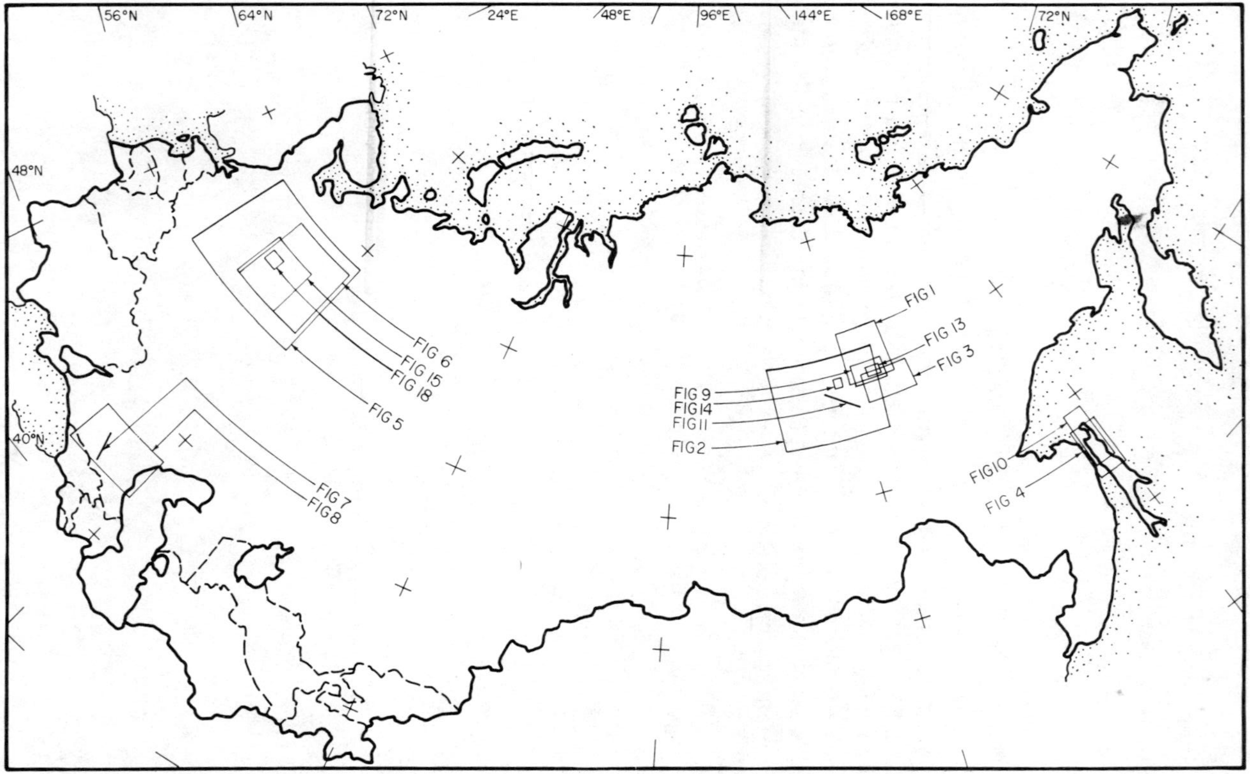

Location map. Numbers refer to figures in text.

INTRODUCTION

The magnetotelluric (MT) method of exploration is widely used as one of an integrated sequence of geologic and geophysical studies in various geologic provinces of the U.S.S.R. The great depths of investigation (up to tens of kilometers), the small size of the field equipment, the simplicity of the method of field acquisition of data, and the relatively low cost of the field work permits the use of MT surveys as a reconnaissance tool in studying regions which are remote and difficult to visit and regions with a great thickness of sedimentary rocks.

Despite the fact that very important geologic results have been obtained using the MT method, the number of publications has been quite limited. The present work aims to correct this situation to some extent.

This book has a secondary objective — the consideration of problems related to applying the magnetotelluric method of exploration in regions with quite different geoelectric conditions. As examples, we have taken the results of the MT investigations carried out primarily to study the geologic structure of several typical regions: the southeastern part of the Siberian platform (in the territory of Western Yakutsk), the Island of Sakhalin, the Moscow syneclise, the Eastern Caucasus forelands, and the Tadjik depression.

TYPICAL GEOELECTRIC SETTING OF SEVERAL AREAS IN THE U.S.S.R. WHICH ARE POTENTIALLY PRODUCTIVE OF OIL AND GAS

Comparison of the geoelectric sequence of various oil- and gas-bearing regions of the U.S.S.R. indicates that in terms of the effectiveness of various types of MT exploration, it is possible to recognize a small number of types of areas:

(1) Areas where the most highly conducting sediments lie direct-

ly on the resistive basement surface; that is, those which form a type *H* sequence (see Glossary). Sections of this type are the most suitable for the use of the MT methods.

(2) Sequences in which the section above basement includes a highly resistive intermediate layer. For specific sets of geoelectric parameters, depending upon the thickness, the resistivity, the depth of burial, the relief of the surface, etc., the presence of an intermediate resistive layer may exert a significant influence on the observation of the electromagnetic field. This influence may vary for several forms of MT exploration. If the effect of this intermediate layer is not considered in interpretation, significant errors in the geologic evaluation of the results can occur.

(3) Sequences in which rocks with moderate resistivity lie between the layers with the highest conductivity and the resistive basement surface. This situation leads to significant difficulties in interpreting data obtained from MT exploration, particularly in cases when the thickness of the rocks with intermediate resistivities is less than the thickness of the overlying conductive beds. Frequently, the presence of the rocks with intermediate resistivities is not apparent on the MT sounding curves.

It is well known that electromagnetic sounding curves reflect more than variations of resistivity with depth. Magnetotelluric sounding curves obtained over a horizontally inhomogeneous sequence of rocks differ to a greater or lesser degree from theoretical curves computed under the assumption of horizontal homogeneity. These differences are commonly called distortions of the MT sounding curve (Dmitriev et al, 1975).

Distortions can be divided into two types based on their characteristics: (1) distortions caused by the development of charge distributions on the surfaces of the inhomogeneities (the galvanic effect); and (2) distortions caused by the effect of circulating currents which arise in the case of an inhomogeneous section (the induction effect).

In regions with a preferred direction of structural trends, the type of distortion depends upon the direction of current flow. The galvanic effect appears primarily on the transverse magnetic (TM or $\rho_{\perp a}$) in which current flow is across the grain of the structure. Several phenomena characteristic of this case are the *S*-effect (a shift of the right-hand portion of the MT sounding curve along the resistivity axis), the screening effect (the washing out of structures above the resistant basement), and the limiting effect (the occurrence of minimums and depressions on the MT sounding curve as a consequence of channeling of currents along the extent of a depression).

Induction effects primarily distort the transverse electric (TE or $\rho\|_a$) in which the current flows along the strike direction of the structure. For such curves, a weak minimum develops characteristically as well as a gradual transition to the rising portion of the curves.

In regions with three-dimensional (3-D) structures, the MT sounding curves are distorted simultaneously by galvanic and induction effects, giving rise to concentrations of currents in depressions and reduction of currents in areas of uplift.

The desire to study the geoelectric section to as great depths as possible, as well as the necessity to recognize effects related to horizontal electrical inhomogeneities, requires one to obtain a complete sounding curve. However, in practice, because of a number of factors, observations are carried out only over a limited range of frequencies. Over this range for specific geoelectric conditions, only a portion of the sounding curve is obtained, leading to yet another problem to be solved in MT exploration. Broadening the frequency range of recording will unquestionably extend the possible applications of the MT methods.

Let us examine the geoelectric conditions in several oil- and gas-bearing regions of the U.S.S.R.

Southeastern part of the Siberian platform (Territory of Yakutia)

Within the limits of Yakutia where studies have been carried out, a number of regional tectonic elements are present. These are the Vilyuis syneclise, the Predpatom boundary basins, the Botuobin region, the Aldan slope, and the Anabar anteclises. A brief description of the geologic structure of each of these regions follows.

In the central portion of the Vilyuis syneclise, neighboring positive and negative structural elements have cubic forms, i.e., the Khapchagi and lower Tung downwarps, and the Linden and Lunkhin uplifts. In the western part of the syneclise, structures are present which are elongate in a north-south direction; these are the Kempendyai and the Vgiattan uplifts with the Suntar basin lying between them.

In the region of the Predpatom boundary basins, the structures are elongate in a north-south direction. Here we recognize the Nui-

Dzherbin and Berezov uplifts, along with lesser structural elements.

The Botuobin area is characterized by cubic structures, namely, the Mirnen Arch and Ichedob-Botuobin downwarp.

The geoelectric sequence across these tectonic elements has considerable variation. In the central part of the Vilyuis syneclise, the geoelectric section grades into a sequence of type H (see Glossary). The high-resistivity basement surface is either the crystalline basement or carbonate rocks of lower Paleozoic age. The electrical properties of the lower Paleozoic rocks which lie at depths of several kilometers are unknown. However, we may assume that because of highly saline connate waters and the existence of fracturing in these rocks, the resistivity may not be very great (possibly less than 10 Ω-m).

Above the resistant basement rocks, terrigenous, possibly highly conductive rocks are present, representing the upper and possibly the middle Paleozoic. The resistivity of this part of the section is no greater than a few ohm-meters, and the thickness possibly as much as 2 to 6 km. The upper, comparatively highly resistive, part of the section consists of terrigenous Mesozoic rocks that contain brackish formation water. The resistivity in these rocks is 20 to 40 Ω-m, while their thickness may be as great as 5 to 6 km. In the surficial part of this sequence, the groundwater may be quite pure, with a resistivity for the rock being 1000 Ω-m or more over a thickness that rarely exceeds 500 m.

A total longitudinal conductance of the sedimentary sequence amounts to several thousand siemens, and the thickness may be greater than 10 km. For these conditions, an MT sounding curve recorded at periods ranging from 15 sec to 1 hour has a descending branch (15–30 $< T <$ 60-100 sec), controlled by the thickness of the uppermost part of the geoelectric section. It then has a minimum and a rising branch ($T >$ 100–200 sec), which characterizes the total longitudinal conductance of the sedimentary sequence.

Low-resistivity Paleozoic rocks are absent from the geoelectric section of the western part of the Vilyuis syneclise, as well as from the northern and southern boundary areas. Carbonate and evaporite Paleozoic beds play an important role in the sedimentary sequence. Their resistivity is tens of ohm-meters, with a thickness amounting to several kilometers.

Total thickness of the sedimentary sequence in the center of the downwarp amounts to 10 to 12 km, while the total longitudinal conductance is about 1000 siemens. An MT sounding curve obtained in these areas is characterized by a left-hand rising segment (10–15 < T < 50–70 sec) reflecting the longitudinal conductance of the section. The right-hand part of the curve (T > 70–100 sec) is characterized by a plateau and a second rising segment which reflect the deeper parts of the earth's crust.

The sedimentary sequence in the Predpatom boundary basin and the Botuobin basin is comprised of relatively highly resistive carbonate and evaporite rocks. The total longitudinal conductance rarely is greater than 100–150 siemens, while in the Botuobin basin, it is less than 20–30 siemens. The thickness of the sedimentary sequence ranges from 2–3 km in the basin to 4–5 km in the boundary basin. The basement consists of highly resistive crystalline rocks of the Archaean basement.

Magnetotelluric sounding curves recorded in the field reflect the longitudinal conductance of the sedimentary sequence in these areas in the form of a left-hand rising segment for a range of periods no greater than 30 to 40 sec. The right-hand part of the MT sounding curve is controlled by the electrical properties of deeper parts of the earth's crust and by distortions from horizontal electrical inhomogeneities in the sedimentary sequence.

An analysis of distortions to MT sounding curves recorded within the Vilyuis syneclise, the Predpatom boundary basin, the Botuobin downwarp, and the Aldan and Anabar anteclises indicates that curves are distorted by a combination of galvanic and induction effects. The strongest distortions are caused by the S-effect and effects accompanying the concentration of currents at the margins between uplifts and basins.

The high resistivity of the permafrost layer has no significant effect on the MT method. There is some distortion to MT sounding curves caused by local inhomogeneities of the permafrost layer in the form of active zones beneath rivers and lakes. The effect of an active zone is to dislocate the MT sounding curve over a wide range of periods. In such areas, the curves cannot be interpreted.

Northern Sakhalin

In northern Sakhalin, the characteristic structure is tight and linear elongated in the north-south direction (along the axis of the island). These are superimposed on more gentle structures, some of which are elongate. (See Figure 4). On the tectonic map, the structures are divided into a central synclinorium and two anticlinoriums bounding it on the east and west. The central synclinorium contains a Cenozoic complex, underlain by Mesozoic and metamorphosed Paleozoic rocks. These same rocks are found in the sequence in the eastern and western anticlinoriums, but the thickness of the Cenozoic complex is much less.

The overall geoelectric section for western and central northern Sakhalin can be divided into three layers comprising a type H section. The base of the section is the horizon of high resistivity attributed to volcanic rocks of Mesozoic or Cenozoic age, or Mesozoic-Paleozoic metamorphic rocks. This is overlain by rocks with low resistivity (4 to 25 Ω-m) representing a sand-shale section with a lithology that varies over wide limits — from lower to middle Miocene to upper Mesozoic. The thickness of these rocks varies from 500 to 5000 m. The upper part of the section consists of relatively resistive beds of Pliocene to upper and middle Miocene age. This part of the section has a thickness of 200 to 5000 m and a resistivity of 20 to 100 Ω-m. The longitudinal conductance of the sedimentary sequence ranges from 50 to 1000 siemens, while the thickness varies from 1 to 10 km.

Magnetotelluric sounding curves recorded over a range of periods from 0.05 to 0.16 sec, as might be expected, clearly reflect a descending branch ($0.1 - 0.5 < T < 0.5 - 10$ sec) characterizing the thickness of the upper layer, a minimum, and a right-hand rising segment ($20–30 < T < 100-200$ sec) reflecting the total longitudinal conductance of the sedimentary sequence. It should be noted that the left-hand portion of these MT sounding curves was obtained over this wide band of periods by using the PMTS equipment, developed by the Institute of Physics of the Earth of the Academy of Sciences. This has greatly expanded the capabilities of MT sounding for solving geologic problems.

Consideration of distortions to MT sounding curves recorded in various tectonic regions of northern Sakhalin leads to the following:

(1) Along the western part of the island, transverse curves are distorted by boundary effects with many curves showing the effect of divergence and concentration of current flow.

(2) In the central part of the island, the primary distortions that appear are those caused by concentration and spreading of current.

(3) In the eastern part of the island, the longitudinal apparent resistivity curve reflects the effect of ocean and local induction effects.

Moscow syneclise

The basement surface in the Moscow syneclise is characterized by a broad structural depression in the northwest-southwest direction, extending from the Baltic Shield on the north to the Voronezh massif and Tokmov arch on the south. Along the axial part of the syneclise, the central Russian allochthon is present (Gordasnikov and Troitskiy, 1966). This structure consists of a series of uplifted and downdropped basement blocks bounded by faults which cause abrupt breaks in the basement surface. The faulting zones, as might be expected, are clearly reflected in the tectonics of the sedimentary sequence.

The sedimentary sequence in the synclinorium is divided into two basic structural-tectonic complexes: the pre-Devonian and the upper Paleozoic. These two complexes are distinguished by different conditions of sedimentation and different styles of structure, and they represent two quite different stages in geologic evolution. The relationship of structural elements in the two complexes is complex. The tectonics of the pre-Devonian beds is clearly related to the morphology of structures on the surface of the basement complex. The tectonics of the upper Paleozoic complex over the axial part of the allochthon usually do not correspond to the tectonic elements in the pre-Devonian complex (positive structural elements correspond to negative elements). Along the margins of the syneclise, there is characteristically more conformity between these two complexes. The overall geoelectric sequence in the Moscow syneclise can be represented as a four layer sequence of type *KH*. The sequence from top to bottom is as follows:

(1) A section consisting of essentially low resistivity (5–15 Ω-m) beds of terrigenous sediments of Tatar age and younger, with a thickness ranging from a few tens of meters in the bordering areas of the syneclise to 400 to 500 m and even more in the central part of the area.

(2) A section of high resistivity (50–150 Ω-m) of beds forming the upper part of the Kazan series to the lower part of beds of the lower Fransk subseries. The thickness of this group of beds increases from west toward the east from 700 to 800 m up to 1500 m and more. These beds are resistive and comprise the basement as seen with direct current sounding methods.

(3) Low-resistivity beds representing the lower Tschigrov and older rocks of the middle Devonian and lower Paleozoic. These are reefs and shoals, the thickness of which varies

markedly from 300 to 1000 m along the borders of the syneclise to 1500 to 2000 m in the central part; thickness increases to 2500 m thickness or more within the limits of the mid-Russian allochthon. The resistivity in this part of the section is 1.5 to 2 Ω-m. It should be noted that reef rocks lying at the base of the lower (terrigenous) complex may have a relatively high resistivity in some parts of the syneclise. In areas where these rocks have significant thickness, the geoelectric section becomes a five-layer section of type *KHA*.

(4) The basement consists of rocks of the Archaean crystalline complex which have high resistivity.

The thickness of the sedimentary sequence ranges from 1600 m along the margins of the syneclise to 4500 m in the axial regions, while the longitudinal conductance varies from 700–800 to 2500–3000 S. The variations in thickness and longitudinal conductance of the sedimentary section result primarily from changes in the third horizon, which are controlled by relief of the basement surface. This provides a favorable set of conditions for mapping the relief of the basement surface in the Moscow syneclise using the MT method.

Magnetotelluric sounding curves recorded over a range of periods from 10 to 900 sec usually consist of a minimum and a right-hand rising segment which reflects the longitudinal conductance of the sedimentary sequence. Analysis of MT sounding curves has indicated that distortions are caused primarily by concentration and spreading effects of current flow. In addition, in zones where the high-resistivity rocks of the second layer in the geoelectric section lie at shallow depths and are strongly faulted, MT sounding curves may show distortions from the *S*-effect and the effect of screening. In these areas, changes in the average direction of the telluric current field are controlled by relief on the surface of the Kazan formation which acts as a screening layer. Where this formaion is deeper, its screening properties disappear and it serves as a cover for the longitudinal components of telluric currents as well as the transverse.

Eastern Caucasus foothills

The eastern Caucasus foothills consist of the Caucasus forelands platform lying between the Precambrian Russian platform and the Alpine folded belt of the Greater Caucasus (see Figure 8). Within the limits of the east Caucasus foothills, the broad Ter-Kum depression is present, which on the south is connected with the basins along the platform margins containing Paleogene and Neogene formations of the Ter-Kaspii basin. The basin consists of broad symmetric structures to the north and tighter structures to the south, separating the platform from the area of Alpine folding. The uplift and the basins are filled with Mesozoic and Cenozoic rocks lying on a tightly folded Paleozoic basement. The surface of the basement is unevenly buried in the south and southeast directions at depths of 1 to 2 km over the Karpii ridge and up to 10 to 12 km in the central part of the Ter-Kaspii basin. The principal structural elements (the Karpii ridge, the East Manichsk basin, the Prikum ridge, and the Ter-Kaspii basin) generally have a subelongate form and are characterized by quite complex structures. Within the boundaries of the mega-anticlinorium of the Greater Caucasus, Paleozoic rocks outcrop on the surface. On the west, the Ter-Kumok basin is bounded by the Stavropol arch, which represents a broad uplift on the Paleozoic basement and is characterized by a marked reduction in the thickness of the sedimentary sequence. A tectonic distortion plays an important role in the formation of the present structure in the Eastern Caucasus forelands; it has caused considerable relief in basement folding and is determined by the block tectonics of this region.

The overall geoelectric sequence in the region is one of type *HA*. Depending upon the relative thicknesses of the individual layers, the resistivity sequence may range from type *H* to type *A* and even to a two-layer type.

The base of the sequence is the faulted and metamorphosed formations of the Paleozoic basement. They are characterized by very high resistivity and form the electrical basement. Immediately above the Paleozoic formations are beds of high resistivity including the beds of the lower chaotic platform complex — Permian, Triassic, and Jurassic rocks. This most thoroughly studied part of the section is characterized by resistivity ranging from a few ohm-meters to tens of ohm-meters or more, and by thickness ranging from zero to several thousand meters. There is little or no velocity or density contrast between the lower part of this sequence and the upper part of the basement. Mesozoic and Paleogene to Neogene beds comprise a layer of low resistivity. The resistivity in the conductive beds is 1 to 3 Ω-m, while the thickness ranges from 1500 to 4000–6000 m. In the upper part of the section, rocks of high resistivity are present (Quaternary and upper Neogene). Resistivity of these varies over wide limits, and the thickness ranges from several tens of meters in the north to several kilometers in the south.

The average resistivity of the sedimentary sequence increases toward the south from 1 to 3 or 4 Ω-m; the most rapid changes occur near the axial part and to the south of the Ter-Kaspii basin. Along the transition from the Ter-Kaspii basin to the mega-anticline of the Greater Caucasus, the thickness of the conductive beds decreases rapidly by a factor of two.

Magnetotelluric sounding curves recorded over a period range from 15 to 2000 sec in areas where the basement is deeply buried have a left-hand descending branch which reflects the thickness of the upper resistive horizons, a minimum, and a right-hand ascending branch which characterizes the total longitudinal conductance of the sedimentary sequence. In areas where the section is not so thick, the curves contain a minimum and a rising branch, and sometimes only a rising branch.

Analysis of distortions to the MT sounding curves indicates that the transverse curves (with current flow across the Greaer Caucasus block) are always distorted by boundary effects, appearing as a decrease in the value for apparent resistivity. The maximum distortion is noted in the southern part of the East Caucasus foothills. With greater distance from the Caucasus mountains and with decreasing periods, the distortions become weaker and the transverse curves become close to the longitudinal. The longitudinal curve (current flow parallel to the Caucasus block) is least distorted and gives accurate information on the vertical extent of the geoelectric sequence.

Tadzhik depression

The Tadzhik depression is an epiplatformal area (intermontane basin) representing a significant break in the course of Alpine orogeny. The sedimentary section in the depression consists of platformal deposits of Mesozoic to Paleogene age and an orogenic molasse of Neogene to Quaternary age. The total thickness of the sedimentary sequence is 8 to 12 km or more.

The principal structural elements have approximately north-south extension. The alternating uplifted and downdropped zones characteristically have an extent ranging from tens of kilometers to several hundred kilometers. Their cross-section is less by a factor of 10 to 15. The sedimentary sequence can be divided into two subsections: that part above the upper Jurassic which lies above the salt horizons, and those beds which lie below the salt horizons. The upper part of the section is a complicated assemblage of fault blocks, formed by a large number of longitudinal and transverse faults. The dip angle in the upper beds is 60 to 70 degrees. The structure in the lower part of this section is relatively unknown, but it may be assumed that it is even more complicated. There are no hypotheses on the structure in the basement or in the lower part of the sedimentary sequence.

The geoelectric relationships of the section in the Tadzhik depression are controlled by a thick conducting sequence, with a resistivity of 3 to 20 Ω-m lying on a practically nonconducting basement surface, with a resistivity of 2000 to 5000 Ω-m. The upper part of the section contains two comparatively high-resistivity (300 to 1000 Ω-m) zones which appear to be the basement for direct current soundings. These are carbonate rocks of the lower Paleocene to upper Miocene and salt beds of the upper Jurassic period. In addition, high resistivities occur in the upper Neogene formations. Between the subsalt Jurassic beds and the basement surface, there are relatively conductive (20 to 30 Ω-m) terrigenous formations of the Permian and Triassic (Koshlakov and Konovlov, 1971).

RESULTS OF THE USE OF THE MAGNETOTELLURIC METHOD IN STUDYING RELIEF OF A RESISTANT BASEMENT SURFACE

Studies of the regional relief of the resistant basement surface (which to a first approximation is the surface of the crystalline basement) is one of the primary applications for MT surveys. Aspects in the solution of such problems include the character and degree of distortion of MT sounding curves, the type of geoelectric sequence, the bandwidth over which an MT sounding curve is obtained, and the amount of information available on the properties of the geoelectric section. Based on the work of Dmitriev et al (1975), we know that the formal interpretation of distorted MT sounding curves (i.e., an interpretation through the use of models with horizontally inhomogeneous media) may give rise to apparently flattened structures, with less character than the actual structure, reflecting the effect of conducting layers. Therefore, it is necessary in compensating for this distortion on MT sounding curves to make use of the least distorted curves caused by horizontal electrical inhomogeneities.

In regions with elongate structures, the most satisfactory information about the properties of the section above the resistant basement (in the absence of edge effects) is given by the transverse curve. Under the same conditions, the longitudinal curve carries useful information about layers that lie beneath the resistant basement surface. In areas with equidimensional block structures, we normally use the average curve which carries information about the layers above the high-resistivity basement. Therefore, the interpretation of MT data begins with an analysis of distortions to the MT sounding curves, then proceeds to the use of the least distorted curves, and later proceeds to the geologic interpretation of these selected curves.

For quantitative interpretation, as one might expect, we use statistical correlations between the various parameters obtained from the MT surveys (total longitudinal conductance, average resistivity, direction of the telluric currents, etc.), and the depth to high-resistivity basement determined from drilling and from other geophysical surveys (Kalenov, 1970). The degree of correlation comprises one of the criteria for believing that the electrical basement is stratigraphically meaningful. With changes in the geoelectric section, the quality of correlation may also vary.

When an adequate frequency range is available from an MT sounding curve with supplementary information on the properties of the geoelectric sequence, various types of simplified and curve matching interpretation may be used. If information on the section is limited, then in many cases an analysis of theoretical curves allows us to set limits on change of thickness and resistivity for individual geoelectric units (Sheinkman and Lipilin, 1968; Yakovlev and Sivitskiy, 1969).

The most satisfactory MT sounding curves for studying the relief on the surface of a high-resistivity basement covered by more conducting rocks are curves which have not only an initial descending segment but also a minimum. When such data are available, for even a small number of reference sites, one can recognize structures with amplitudes of only 8 to 12 percent of the total thickness of the conducting section. With the presence of intermediate zones of high resistivity in the conducting sequence, lying directly on the high resistivity basement, the problem of mapping the surface of the basement is more complicated. Especially unfavorable conditions occur when the thickness of a layer with high resistivity is no greater than the thickness of the overlying more conductive beds, since under these conditions, the presence of the higher resistivity beds cannot be recognized on the MT sounding curves.

Classifications of areas with respect to the type of geoelectric sequence and the degree of correlation between depth to the basement surface, longitudinal conductance, and average resistivity represent one quite widely used means for evaluating or recognizing serious errors (many amount to several tens of percent) in interpreting MT survey data. Let us consider examples of several types.

Southeast part of the Siberian platform (Yakutsk Autonomous Republic)

The deep geologic structure in the Yakutsk region has been studied primarily using geophysical methods with a limited amount of information provided by deep drilling. The area has been covered with gravity surveys and aeromagnetic surveys, giving a first approximation to the regional tectonics. Seismic refraction shooting has provided information about the thickness of the sedimentary sequence. Reflection seismic shooting has been used primarily for mapping local structures within the sedimentary sequence.

Magnetotelluric surveying to provide information on geologic structure in the Yakutsk area has been carried on since 1958 (Yakovliev, 1964). Surveys were carried out over the principal structures in the area on scales of 1:1,000,000, 1:5,000,000, and 1:200,000, covering a total area of approximately 500,000 km^2.

An analysis of these MT soundings has indicated that the most satisfactory geologic information is provided by (1) the effective curves (i.e., with the two principal sets of apparent resistivity values being averaged) obtained near block-faulted structures and in basins; (2) longitudinal curves obtained over the central parts of uplifts; and (3) transverse curves obtained near the edges of elongated uplifts.

The effectiveness of MT sounding curves was investigated in this respect over the equidimensional block structures of the central part of the Vilyuis syneclise. In the western part of the Vilyuis syneclise, where elongated structures are present, MT longitudinal sounding curves obtained over the axial parts of the uplift, and

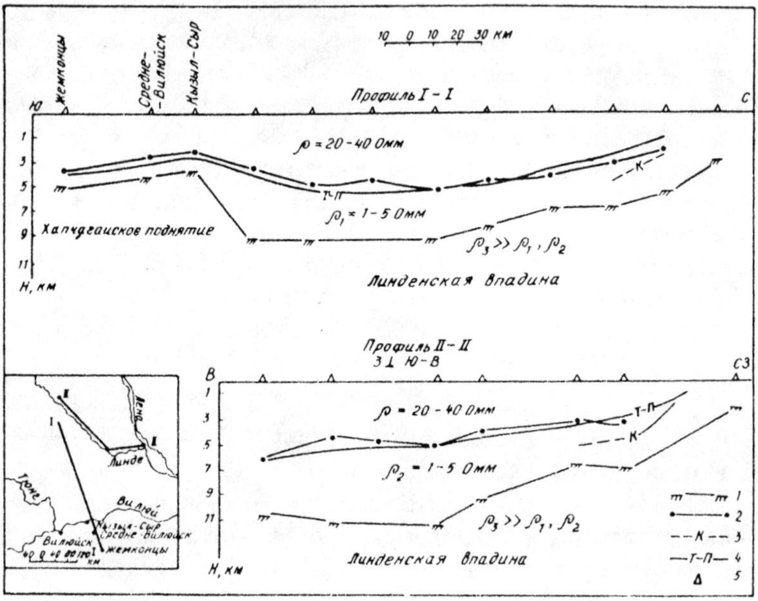

FIG. 1. Results of MT profiling done across the Kapchagai uplift and Linden basin (Yakutia). (1) Top of the high-resistivity beds. (2) Top of the low-resistivity beds. (3) Seismic horizon *K*. (4) Seismic horizon *T-P*. (5) Locations of MT soundings.

transverse curves obtained along the edges of the uplift, were interpreted. In the Predpatom basin (characterized by structures with considerable linear extent) and in the Botuobin basin (where structures are more or less equidimensional), transverse and effective MT sounding curves were found to be more useful. Exceptions include the axial zone of the Angaro-Len and Nuis-Dzherbin basins where interpretation was based on longitudinal curves.

The interpretation based on effective MT sounding curves along profiles 1-1 and 2-2, which cross the central part of the Vilyuis syneclise, provided information on the depth to the high-resistivity basement or to the carbonate beds of lower Paleozoic age immediately above it (see Figure 1). Depths were determined from the product of longitudinal conductance in the sedimentary sequence and the average longitudinal resistivity which was determined by the method described by Berdichevskiy (1968).

In the Khapchagai basin, the resistive basement lies at a depth to 5 to 6 km, deepening in the Linden basin to 9 to 10 km along profile 1-1 and up to 11 to 12 km along profile 2-2. The northern edge of the Linden basin is formed by a rise in the resistive basement surface with depth decreasing to 2 to 3 km. Characteristically, the depth to the resistive basement is significantly greater than the depth to the seismic horizon "K" which correlates with the surface of lower Paleozoic carbonate formations. This is evidence that the high-resistivity basement surface corresponds to the crystalline basement.

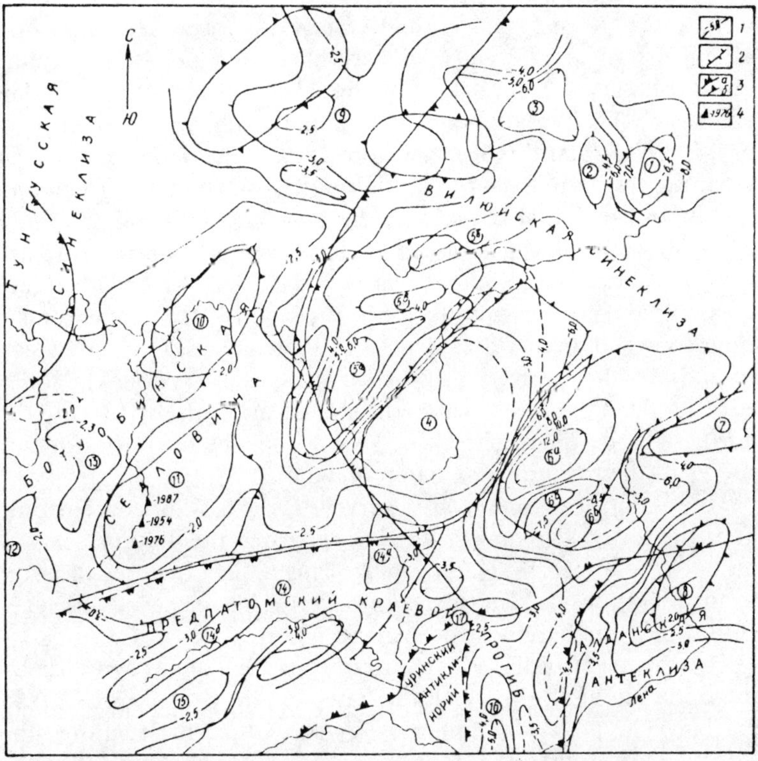

FIG. 2. Structural relief on the basement surface in southwest Yakutia based on MT data (from Yakovlev and Bubnov). (1) Contours of the basement surface (kilometers). (2) Fault zone. (3) Boundaries of tectonic elements (a) major; (b) first and second order. (4) Wells to basement, with the depth to basement (meters).

From data provided by the interpretation of MT sounding curves obtained in the northwestern part of Yakutsk, a structural relief map on the basement surface has been prepared (Yakovlev and Sivitskiy, 1970). The depth to basement is determined from curves for *S(H)*, where *S* is the longitudinal conductance of the sedimentary sequence and *H* is the depth to the surface of the basement. Values for *H* are provided by drilling data in the Botuobin basin, on the northwest side of the Aldan anticline, and from seismic reflection data in the Kempendyai basin and the Botuobin basin (see Figure 2).

Within the Vilyuis syneclise, structures of first order which may be recognized are the Kempendai (6)* and Vgiattan (5) basins and the intervening Suntar (4) uplift. The basement lies at depths less than 2 km on the Suntar uplift, at 6 to 7 km in the Vgiattan depression, and at 10 to 12 km in the Kempendyai basin. The basins are asymmetric structures with the steep limb lying against the Suntar uplift and the gentle limb lying on the side of the Anabar and Aldan anteclises. The Vgiattan basin is complicated by the relatively broad Vilyuis-Markin (5b) shelf, present at a depth of 4 km, which separates the basin from the south Vgiattan (5a) and north Vgiattan (5b) basins. A similar structure is the Arbi (6b) platform, present in the Kempendyai basin at a depth of about 8 km, which separates basins in the central Kempendyai (6a) and south Kempendyai (6c) areas. To the northeast in the Kempendyai basin, the upper Sin (7) ridge forms an uplift with a depth to basement of 3.5 to 4 km which complicates the south edge of the Vilyuis syneclise. Along the south edge of the Vilyuis syneclise, the Aldan anticline is present, the northern slope of which can be seen within the territories of this survey in the form of a namanin downwarp.

To the northeast of the area surveyed, the 6.5-km contour outlines the lower Tungus (1) downwarp, while the 4.4-km contour outlines the upper Tyalichim uplift. This uplift separates the west Syangdin (3) basin from the northwestern edge of the Vilyuis syneclise, where the basement deepens to 6 km.

The Vilyuis syneclise is separated from the Tungus syneclise by the Botuobin arch, on which the basement lies at depths of 2.0 to 2.5 km. The arch is complicated by second-order structures: the Mirnin (11) dome, the Ichodo-Botuobin downwarp (12), the Sildukar (10) ridge-like uplift, and the Dzunkun basin (13). The Mirnin dome and the Sildukar uplift, according to MT sounding and profiling data, are local basement steps. These are reflected in sedimentary section by uplifts or anticlines that appear in the

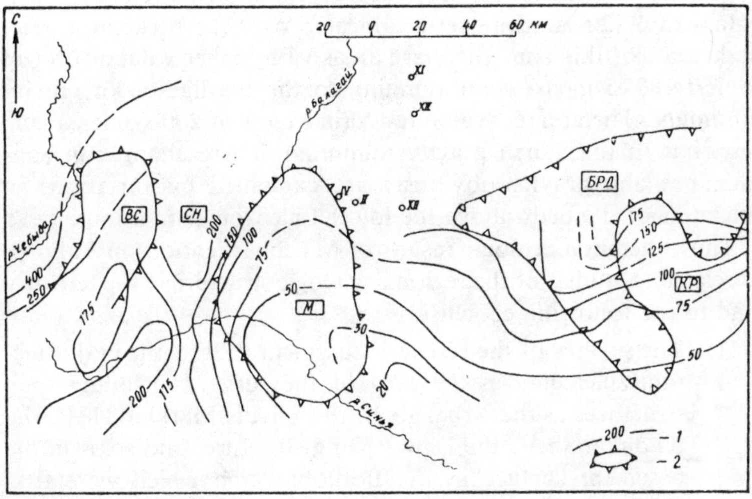

FIG. 3. Map of the conductance of beds above the lower Paleozoic (south edge of the Vilyuis synclise). (1) Conductance contour. (2) Gravity anomaly. Gravity maximums: BC-Upper Sin; M-Meklelin; KP-Kiramdin. Gravity lows: CH-Sanan; BPD-Berdigestyakh.

seismic reflection sections. Within the northern part of the Botuobin arch there is a sharply defined basin separating it from the southern slope of the Anabar anteclise. Within the area surveyed, this limb of the anteclise is complicated by the Morkokin (9) downwarp to the basement, where the total thickness of the sedimentary section is only 2.5 to 3 km.

Along the south side of the Botuobin arch, a fault zone separates it from the structure of the Predpatom boundary basin. The deepest of these basins is the Nyui-Dzherbin basin (14) where the thickness of the sedimentary section is 4 km. The basin is complicated by the Vilyuis-Dzherbin (14a) and Peledui (15) uplift and Nyuisk basin (14b).

The Predpatom basin is also related to the Berezov (16) basin with a depth greater than 5 km and Biryuk (17) syncline, where the basement lies at a depth of about 2.5 km.

Along the south edge of the Vilyuis syneclise, MT exploration was used to study relief on the surface of resistant Paleozoic formations comprising the lower part of the sedimentary sequence. A map showing the variation of longitudinal conductance of the beds above these resistant rocks is shown in Figure 3. On this map, anomalies in gravity are also noted. Zones with relatively low values of *S* (20–50 siemens) are located over the Mekelin gravity high. East of this zone there are areas with higher values of *S* (up to 160–180 siemens) corresponding to the Berdigestyakh gravity minimum. The greatest value for *S* (in excess of 200 siemens) corresponds to the Sanan gravity minimum. These changes in conductance and gravity obviously are explained by variations in thickness of the beds above the lower Paleozoic formations.

Thus, the basic geologic results of MT investigations in Yakutsk as related to studies of the regional geologic structure in the territory lead to the following conclusions:

(1) The surface of the resistant basement reflects the major tectonic elements in the area of the survey, including such structures as the Arbai shelf, the Vilyuis-Markhin shelf, the Suldukar shelf, the Lower Tung structure, and steps in the basement surface in the Botuobin arch which were first recognized from MT data.

(2) The first satisfactory determinations of the thickness of the sedimentary section were made in the downdropped areas of the Viliosk syneclise: the Linden, Vgiattan, and Kempendyai basins.

Data provided by the regional MT network in many cases permitted a more systematic planning for subsequent detailed seismic reflection surveys.

Northern Sakhalin

Early ideas about the tectonic setting in northern Sakhalin were obtained from a regional gravity survey and from deep drilling. Over much of the territory of northern Sakhalin, only the upper part of the sedimentary sequence (up to 3.5 km) has yet been studied, and deeper horizons have been penetrated by only a few wells. Seismic refraction studies have encountered unsatisfactory operating conditions, and it has been possible to determine the thickness of the sedimentary sequence only in the northwest part of northern Sakhalin.

Magnetotelluric sounding surveys were undertaken in northern Sakhalin in 1964 to assist in determining the thickness of the sedimentary sequence. Between 1964 and 1972, the survey covered an area greater than 30,000 km^2 at a scale to 1:1,000,000. From a study of the results that have been obtained from analyzing the MT sounding curves, it has been noted that the least distorted curves

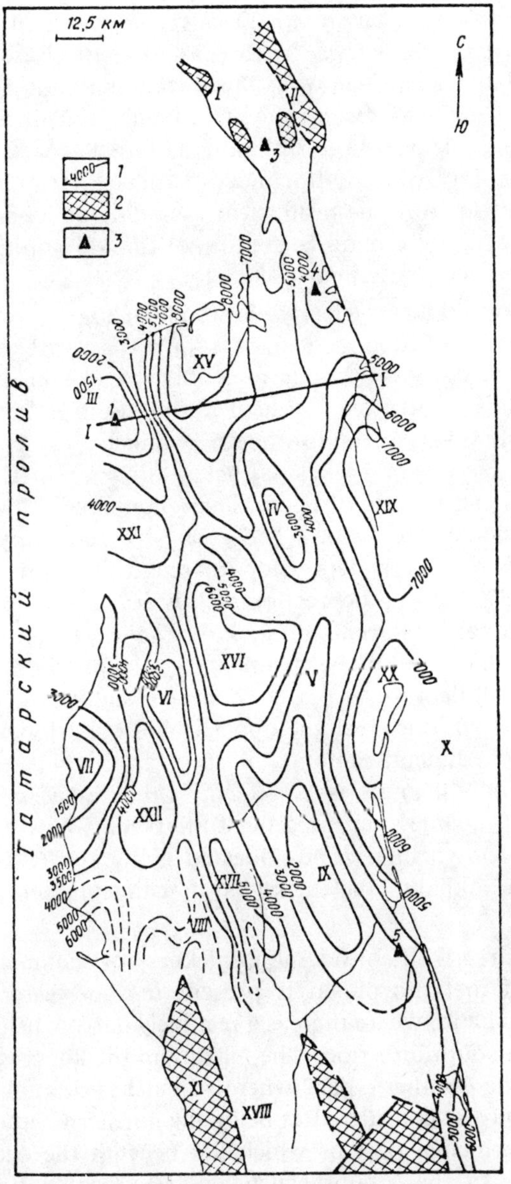

FIG. 4. Map of the thickness of the sedimentary section in North Sakhalin. (1) Isopach of the sedimentary rocks (meters). (2) Areas of outcrop of pre-Neogene. (3) Wells penetrating to the pre-Neogene.

for use in interpretation are the longitudinal curves for the western part of the island, transverse and effective curves in the central part, and transverse curves in the eastern part. Based on these sounding data, a map showing the thickness of the sedimentary section has been prepared, as shown in Figure 4. This map reflects relief on the surface of a resistant basement (Alperovich and Chernyavskiy, 1973). The thickness of the sedimentary sequence was determined from the product of longitudinal conductance, as specified by the right-hand asymptote of the MT sounding curves and the average longitudinal resistivity.

The map in Figure 4 clearly shows all of the major structural elements that have been recognized using a variety of geologic and geophysical methods. Within the area of northern Sakhalin there are characteristically very large thickness variations. The maximum thickness of weakly metamorphosed sedimentary rocks is found within the major graben-like Nishsko-Baikal basin (XV–XVII), which represents a zone of depression lying between two external interdepression horst blocks. The greatest sedimentary thickness found within the Baikal depression (up to 9 to 10 km) is apparently related with the strong depression of this area in the pre-Neogene and possibly early Miocene times. To the south, there is a marked decrease in thickness of the sedimentary section to 3.5–4 km and within the Val depression one sees another significant increase in thickness to 7–8 km. Over the upper Dagin horst block (IX), the thickness of sediments amounts to 3 km. In the Nish depression, the maximum thickness is 5.5–6 km. Other areas of increased sedimentary thickness are present at the Nabil-Chaivin (XX) and Pil-tun (XIX) synclinal zones, located along the eastern shores of northern Sakhalin; the thickness of sediments here ranges up to 7–8 km.

Another area with considerable thickness of sedimentary rocks and limited metamorphism is present in the western part of northern Sakhalin, where there is a relatively narrow belt elongated in the north-south direction. The maximum thickness in this zone occurs in the southern part where it reaches 7 km. This area obviously may be considered as being the northern end of a major basin, the principal part of which lies beneath the ocean in the Tatar Gulf. In the Vagios syncline (XII), located between the Engizpal (VI) and Pogibin (VII) uplifts, the resistant basement lies at a depth of about 5 km. Farther to the north, the thickness of the sedimentary sequence thins to 4 or even to 3.5 km, but within the Chingai basin (XXI), (the most northerly part of the area under consideration), the sedimentary thickness increases again up to 5 or 6 km.

To the west and to the north of the Nish-Baikal synclinal zone, within the Laungri (III), the Engizpal, and the Dangin-Girgilanin (IV-IX) uplifts, the thickness of the sedimentary section thins markedly to between 1 and 3.5 km.

The results from MT sounding are in good agreement with data from gravity studies, refraction surveys (in the northwest part of the island), and seismic reflection surveys. In particular seismic reflection shooting done in the Sea of Okhotsk (in the northern part of the Baikal-Pomor synclinal zone) shows a high-velocity horizon (V = 6.3–6.4 km/sec) at a depth of about 13 km; this has been assigned to the surface of the metamorphic complex and thus does not contradict the MT sounding data. A high-velocity intermediate boundary with V = 5.4–5.6 km/sec on the marine deep seismic profile obtained across the central part of the Nish Depression was detected at a depth of about 6 km. This is the thickness of sediments obtained by interpreting MT sounding data.

The zones of maximum and minimum sediment thickness recognized with MT data in the northern Sakhalin survey area are in good agreement with regional gravity anomalies having the appropriate sign. Consequently, the results of a regional MT survey carried out over northern Sakhalin were used to prepare the first map showing the thickness of the Paleogene-Neogene sedimentary section. This is now used for comparative evaluation of the prospects of the occurrence of oil and gas in various parts and for organizing subsequent oil and gas exploration operations.

Moscow syneclise

The geologic structure in the Moscow syneclise has been studied using geophysical methods and drilling. A gravity survey and a magnetic survey at a scale of 1:500,000 were used to obtain the first quantitative description of the regional structure on the basement. Seismic refraction surveys carried out along a number of profiles provided some information about the thickness of the sedimentary section, and seismic reflection surveys have been used to map a number of structures within the sedimentary section. A few deep wells have provided stratigraphic information about the section in the syneclise and data for use in interpretations of the various geophysical surveys.

An MT survey was undertaken in 1962 for the purpose of study-

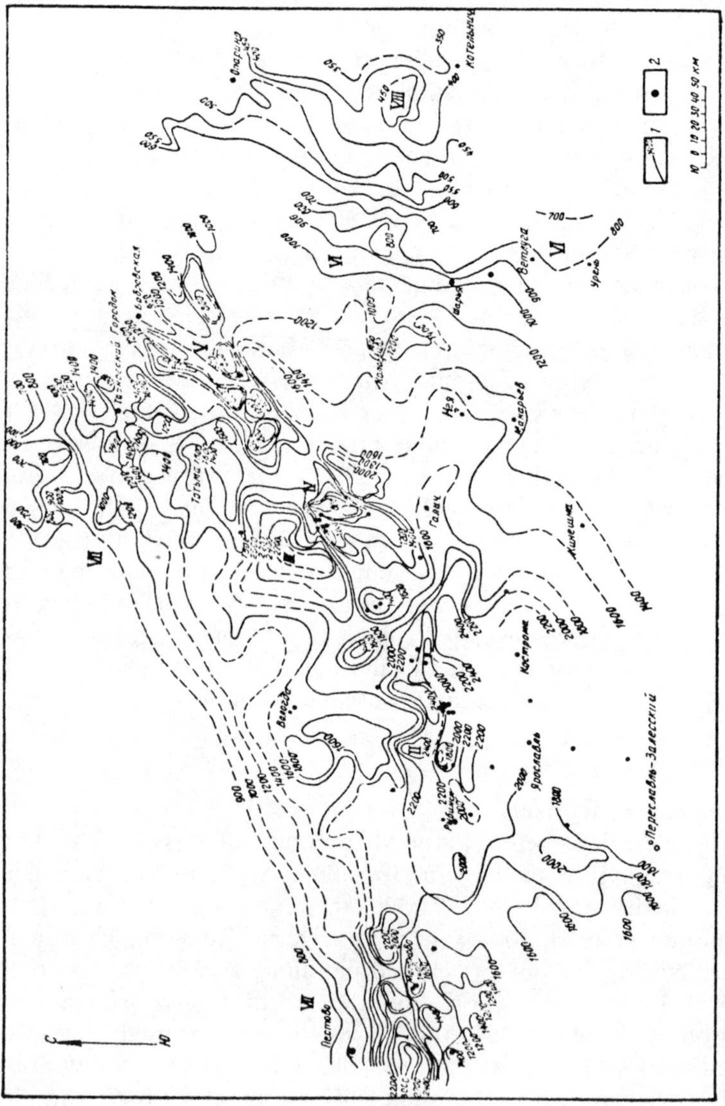

FIG. 5. Composite map of the total conductance in the Moscow synclise (compiled by A. L. Sheinkman). (1) S contours (siemens). (2) Deep well.

ing the relief of the basement surface within the Moscow syneclise. A network of MT profiling was undertaken at scales of 1:200,000 and 1:500,000 over an area of about 250,000 km^2; in addition, about 400 MT soundings were completed.

In regional studies of the relief on the basement surface, primarily the effective values for the MT parameters were used, due to the block faulted structure of the basement.

A composite map of values of longitudinal conductance obtained from MT sounding, MT profiling, and frequency-domain EM sounding was prepared at a scale of 1:500,000 for the entire part of the Moscow syneclise surveyed; this map reflects the regional structure on the basement surface in the syneclise (see Figure 5). The area with the greatest longitudinal conductance is located over the axial part of the middle Russian allochthon. Within the Valdai basin (I), the value amounts to 2700–3000 S, in the Pomekhon (II) basin, it is 2400 S, and in the Biryakov depression (III), it is 2700 S. The thickness of the sedimentary sequence in these areas ranges from 3.5 to 4.5 km, while the average resistivity ranges from 1.4 to 1.7 Ω-m.

To the southeast of the axial zone of the allochthon the value for S decreases from 2000 to 800 S, clearly defining the eastern boundary of the syneclise (VI). In this same direction, the depth to basement also decreases from 3.3 to 2.3 km, and average resistivity increases to 3 Ω-m.

To the east, within the Kotel'nich arch (VIII), the conductance also decreases, to 800 S, because of the rise of the basement surface toward the edge of the syneclise to depths of 1.8–2.0 km. The resistivity in this area is 2.4–2.8 Ω-m. The 900-S line traces the northern edge of the Moscow syneclise over a distance of 500 km (VII). It should be noted that in the Coligalich area (IV) and the Roslyatin (V) grabens, related to the axial zone of the mid-Russian allochthon, the value for S is significantly lower (range 800-1000 S) as a consequence of the screening of the telluric field by the resistant interval (II) in the geoelectric section. In this region, the anomaly indicates primarily the relief from the surface of the high-resistivity Kazan formation.

Thus, the map of longitudinal conductance over most of the Moscow syneclise can be used quantitatively in interpretation as a map showing changes in the total longitudinal total thickness of the sedimentary section. This map provides a general representation of the principal features of the deep geologic structures in the syneclise.

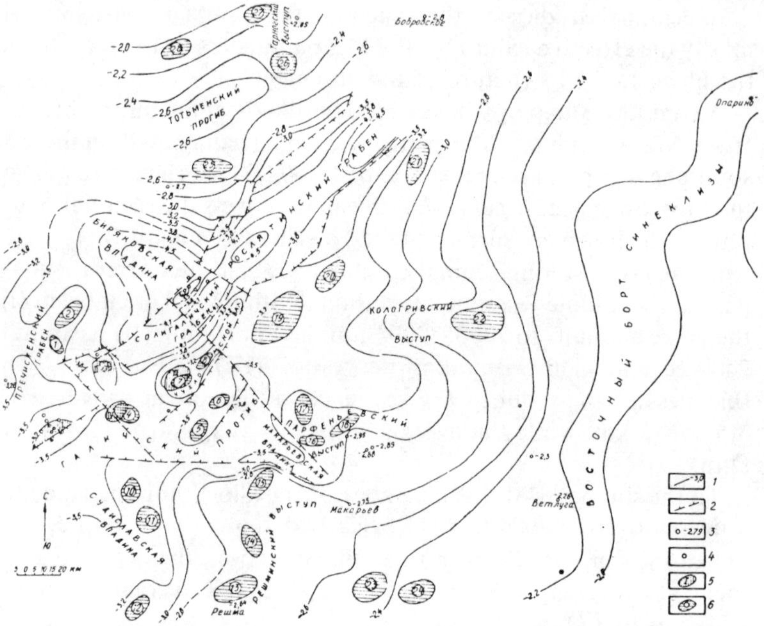

FIG. 6. Structural map of the basement surface in the Moscow synclise based on electrical exploration (compiled by A. L. Sheinkman). (1) Depth to basement (kilometers). (2) Tectonic disruptions. (3) Wells and absolute depths to basement. (4) Depths of wells not penetrating basement. (5) Structural elevations based on seismic reflection, refraction, and drilling: I-Lubim, II-Dyakonov, III-Orekhov, IV-Buis. (6) Structural elevations of the basement and lower part of the sedimentary section based on electrical exploration: 1, Ramen; 2, Vosin; 3, Dyakonov; 4, Buis; 5, Galich; 6, Burmakin; 7, Orekhov; 8, Chukhlom, 9, Soligalich; 10, Susanin; 11, Sudislav; 12, Privolsh; 13, Peshmin; 14, Yurov; 15, Filin; 16, Losin; 17, Parfenyev; 18, Neis; 19, Sudais; 20, Kop'ev; 21, Dubrov; 22, Pishug; 23, Naumov; 24, Ilin-Zabor; 25, Nefedikhin; 26, Fedorov; 27, Okhulov; 28, Shlichev.

Over most of the area of the Moscow syneclise, a broad scale (1:200,000) telluric network was completed, supplementing the MT sounding network. A map of the telluric voltage *E* averaged over a coarse grid with a radius of 12 km was prepared. The result of spatial filtering of 12-km wavelengths clearly reflects regional variations in the conductance of the sedimentary section, so this map was converted to a schematic map showing relief of the basement surface. Data were obtained quantitatively using interpretation charts constructed from an analysis of theoretical four-layer type *KH* MT sounding curves (Sheinkman and Lipilin, 1968). These interpretation charts were used to determine values for longitudinal resistivity, and to determine depth (*H*) to basement using a relationship between the spatially averaged electric field and *H*. Use of this procedure enabled us to divide the area into a number of subareas such that within each subarea, average resistivity was approximately constant. Thus, values for *E* were closely related to variations in thickness of the sedimentary section. The area of the Soligalich and Roslyatin grabens mentioned above, where the telluric field is not related to the depth to basement, is excluded. In these structures, the thickness of the sedimentary section could only be determined at the specific MT sounding sites.

After the region had been subdivided, a map of the relief of the crystalline basement was constructed as shown in Figure 6. This clearly showed the northern and eastern borders of the syneclise, and allowed a number of structures to be recognized in the central area.

Along the northern boundary are located the Tarnog step fault in the basement (oriented north-south), and the broad Tot'men basin. The latter apparently provides the monoclinal edge of the Viryakov uplift.

The eastern border of the syneclise is characterized by contours on the basement surface oriented approximately north-south. In the east is the Kotel'Niches dome and, at the edge of the syneclise, the Kologriv step fault in the basement.

The area including the southeast edge of the syneclise contains the Reshmin and Parfen'ev basement uplifts which define the Makarev basin. To the east, the Reshmin uplift borders the Sudislav basin as well as the Makarev basin, which together define the limits of the Galich basin, in the central part of the syneclise. The Galich basin in the mid-Russian allochthon is characterized by very broad structure and is controlled by northeast striking fault.

Within the mid-Russian allochthon there are a number of

Table 1.

Area and well identification	**Depth to basement (in km) From drilling**	**Electrical data**	**Difference (percent)**
Dyakonov-1	3.13	3.15	0.5
Dyakonov-3	3.24	3.2	1.0
Galich	3.8	3.4	11.0
Neya-1	3.1	3.2	3.0
Neya-3	3.0	3.15	5.0
Rozhdestvin	2.45	2.65	8.0
Poshekhon	2.94	2.95	0
Tot'ma	2.79	2.8	0
Lezha	3.49	3.65	4.5
Tarnog city	2.19	2.35	7.0
Orekhov	3.45	3.4	11.5

graben-like basins which form an overall synclincal structure on the surface of the basement. In the west is located the Prechisten graben, striking north to northeast. The broad Soligalich graben is in the Chukhlom-Soligalich region, with its central and deepest part transecting the tectonic pattern in a north-south direction. North of the Soligalich graben is the Biryakov uplift, and to the east of the allochthon one finds the north-south elongated Roslyatin graben, which is controlled by basement faults.

The results from MT exploration, which have permitted determination of the depth of the basic tectonic elements of the crystalline basement in the Moscow syneclise, were confirmed through deep drilling. Eleven wells were drilled in areas with varying geoelectric character to depths penetrating the crystalline basement. Table 1, which presents data from both MT exploration and drilling results, was recently prepared. It is obvious that in eight cases, the two depths to basement are practically the same.

Results of investigations in the Moscow syneclise (Table 1) made it possible to establish the relief on its surface in the central and eastern parts. This demonstrates the high degree of geologic effectiveness of MT exploration. This application provided a very satisfactory solution to questions about the regional investigations in the area surveyed.

Eastern Caucasus foothills

Geologic studies in the eastern Caucasus foothills have been carried out for many years. At the beginning of the 1960s, these studies, along with seismic reflection surveys carried out over a survey network which covered the entire area, provided a reasonably complete picture of the structure of the section to the base of the Mesozoic. Deep boreholes penetrated basement only in a few areas having limited thickness of sedimentary rock. Therefore, an integrated program of geophysical investigations, including seismic refraction, electrical exploration, and gravity surveys, was carried out to study the relief of the basement and the lower part of the sedimentary sequence.

The MT studies which were part of this integrated approach were carried out in the eastern Caucasus foothills starting in 1962. Both profile and aerial studies at a scale of 1:500,000 were carried out. The length of the profiles was about 2500 km.

Study of the MT sounding curves indicated that within the Ter-Kaspii basin and the southern flank of the Prikum rise, the least distorted and most satisfactory curves for use in quantitative interpretation were the longitudinal MT sounding curves, whereas over the central part of the Prikum rise, the eastern Manich basin, and the Karpin rise, longitudinal, transverse, and effective MT sounding curves were used.

In areas with relatively shallow basement, where the electrical section appeared to be two layers or transitional to type H, the thickness of the conductive beds was determined from a relationship given by Berdichevskiy (1968):

$$H = S\,\rho_{T,\,\min}\,P,$$

where S is the total longitudinal conductance, $\rho_{T,\min}$ is the value of apparent resistivity at the minimum on the MT sounding curve, and P is an arbitrary parameter. The conductance and the minimum apparent resistivity are taken from the MT sounding curve, while the parameter P is determined either from a nomograph given by Berdichevskiy (1968) or by considering drilling data and electric logs. The depth to the surface of the basement obtained in this way is virtually the same as the depth obtained from refraction data and drilling (see Figure 7). In zones with deeper sections (geoelectric sections of type A or HA), the thickness of the sedimentary sequence was determined using a correlation relationship

$$S = f(H), \text{ or } S\rho_{T,\,\min} = f(H),$$

where H is the depth to basement obtained from refraction data or drilling (Averbukh, 1973). The correlation relationship that was obtained was nearly linear and exhibited a high correlation coefficient, greater than 0.9. With the transition from one tectonic

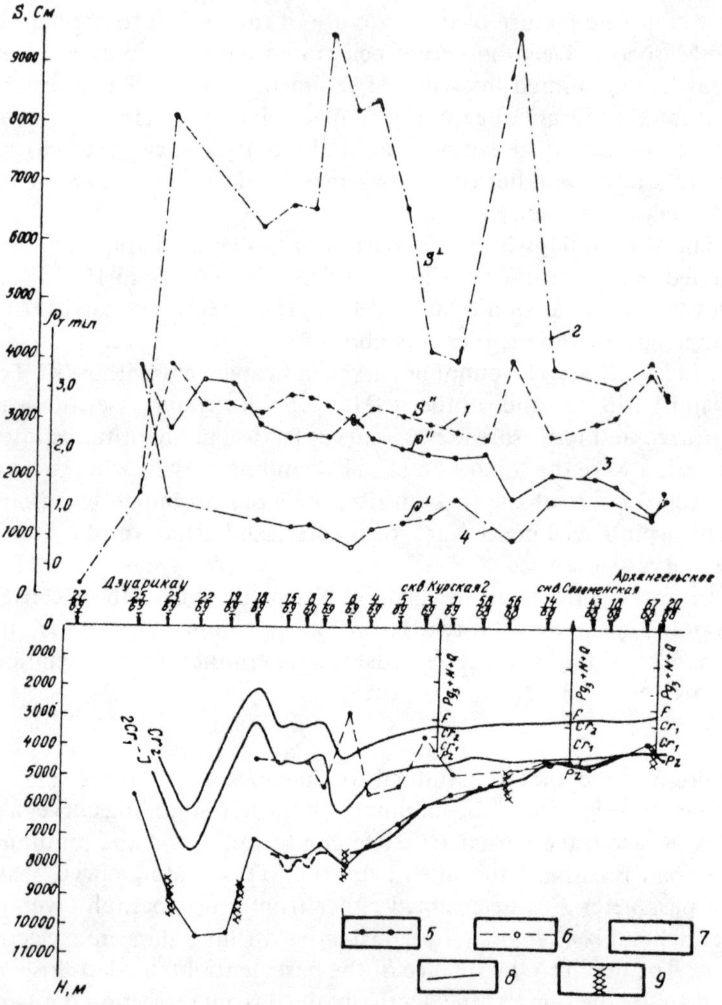

FIG. 7. Geologic and geophysical profiles from Arkhangel to Dzuarik in the eastern Caucasus foothills. (1) Profile of conductance from longitudinal resistivity. (2) Profile of conductance from transverse resistivity. (3) Profile of minimum longitudinal resistivity. (4) Profile of minimum transverse resistivity. (5) Surface of the electrical basement from MT data (longitudinal resistivity). (6) Surface of the electrical basement from MT data (transverse resistivity). (7) Refracting horizons from seismic refraction survey. (8) Reflection horizons from seismic reflection survey. (9) Fault zone.

setting to another, and consequently with a change in the geoelectric section, the regression coefficient varied, requiring that the data be subdivided into consistent groups based on the type of section. The error in determining depth to basement using this correlation approach did not exceed 10 percent provided specific tectonic settings were considered.

In some regions with geoelectric sections of type *A* or *HA*, drilling and detailed seismic reflection surveying are sufficient such that the upper part of the sedimentary sequence is well known, at least to the top of relatively resistant formations (corresponding to the reflecting horizon 2K 1-J which rest on the basement surface). For such cases, the thickness of the sedimentary section was calculated using the expression

$$H = H_1 + \rho_{pr}\left(S - \frac{H_1}{\rho_1}\right),$$

where H_1 and ρ_1 are the thickness and average resistivity of the formations above the 2K 1-J horizon, and ρ_{pr} is the average resistivity of the resistant formations on the basement surface as determined at reference points from MT sounding data and from drilling (Konnov and Varlamov, 1971; Averbuch, 1974).

A useful property of type *A* or *HA* curves is that in cases where the thickness of the conducting formations is sufficiently large, the value for ρ_T at the minimum on the MT sounding curve is nearly resistivity and thus approximately equals $\rho_{T,\,\min}$ for a curve in which the intermediate layer is absent. Consequently, if the parameters for the upper part of the section are known, it is possible to determine the average resistivity of these formations from $\rho_{T,\,\min}$ (Berdichevskiy, 1965). We used this property in specific cases where the thickness of the intermediate layer of high resistivity was no greater than the thickness of the more conducting formations. In the east Manich basin, the surface of the electrical basement constructed by this method was close to the probable surface of the crystalline basement, with disagreement between electrical data and refraction data being 250 m. This degree of accuracy is completely in accord with the accuracy of the basic correlation used in the analysis.

From the interpretation of MT data in combination with the various other geophysical methods (seismic refraction and gravity) and drilling, a schematic map of the relief of the basement surface was prepared, as shown in Figure 8 (Averbuch, 1974). In preparing this map, electrical exploration data were utilized for determining the depth to basement between refraction profiles and in areas where refraction surveys were not carried out (principally in the southern part of the area). This approach reflects the fact that in comparison with boundaries within the sedimentary sequence, the surface of the basement is the most significant boundary that could be differentiated. The various major tectonic features in the area mapped could be clearly recognized: the Karpin rise, the East Manich basin, the Prikun rise, the Chernoless basin, and the Ter-Kaspii basin.

In general, the basement surface is characterized by progressively deeper burial to the south and southeast. In the north, the minimum thickness of sediments (1.5–2.5 km) is found over the Karpin

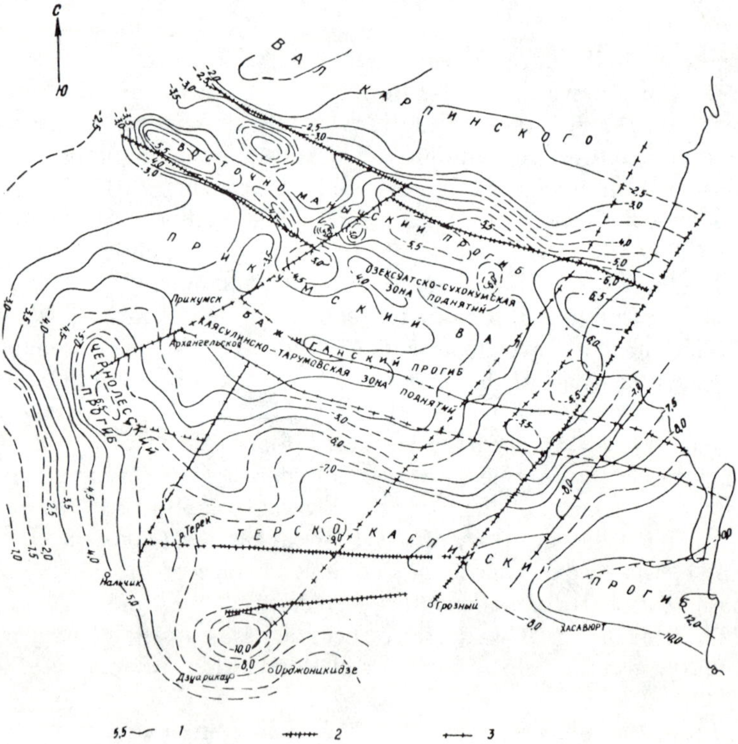

FIG. 8. Structure map on the basement in the eastern Caucasus foothills (from geophysical data and drilling). (1) Contour on the basement surface (kilometers). (2) Fault in the basement or lower part of the sedimentary section. (3) Basement fault.

rise. To the south, the basement slopes below the east Manich basin. Depth increases gradually at first, but then rapidly, to reach a maximum thickness of 5.5–5.6 km. The general structure of the depressed area is complicated by a number of relative uplifts and downdrops in portions of the basement more marked at the north edge than at the south edge.

The south slope of the Prikun rise is relatively shallow and grades gradually into the Ter-Kaspii shallow basin. Over the axial part of the basins, the depth of the basement increases toward the east from 6 to 12 km. The horst-like uplifted zone oriented approximately north-south separates the basin from several smaller basins. To the northwest of the Ter-Kaspii basin, across the intervening horst, one finds the Chernoless basin which has an axis elongated in the northwest-southeast direction. This is a down-warp with an asymmetric structure (the southwest limb being deeper than the northeast limb), and it has a maximum thickness of sediments of 6500m on its axis. To the south of the axial part of the Ter-Kaspii basin, the basement rises abruptly into the greater Caucasus anticlinorium.

These various studies provided an overall characterization of the primary features of the regional geologic structure of the eastern Caucasus foothills, i.e., evidence on the relief of the metamorphic basement, which has considerable value in evaluating the prospects for oil and gas occurrence in this region. As can be seen from the results that have been given, the MT method can be used in a study of the relief of resistant rocks, which ordinarily would be the crystalline basement over a wide variety of geologic conditions. In addition to regions in which we have discussed results, similar results were obtained in studies of the crystalline basement in the fringe areas of the Dneprov-Donetz basin (Tomchakov and Kharchenko, 1976), the southern part of White Russia (Stadnik et al, 1976), the Timan-Pechor province (Podlovilin and Gablin, 1976), and the Tunguess syneclise (Sharporev and Bubnov, 1976). Resistive formations of various ages have been the target of research studies in west Siberia (Kopelev et al, 1976), Azerbaijan (Abdullaev et al, 1974; Faradshev et al, 1976), Kamchatka, Uzbekistan, and the Dneprov-Donetz basin (Tomchakov and Karchenko, 1976), and in a number of other areas within the U.S.S.R. These studies have indicated the feasibility of mapping high-resistivity formations of various ages in the Cis-Caspian basin (Brusyanin et al, 1976; Sidorov et al, 1976), Georgia (Gugunava et al, 1976), and Turkmenia regions (Dubrovskiy, 1976).

FEASIBILITY OF SUBDIVIDING THE SEDIMENTARY SEQUENCE ON THE BASIS OF MAGNETOTELLURIC EXPLORATION

Conditions which are favorable for the subdivision of the geoelectric section using MT methods are: (1) the presence of thick intervals with characteristic resistivities, (2) the ability to record the portion of the MT sounding curve over which such thick sequences exert the strongest effect, (3) the presence in the section of intermediate screening layers which are transparent to one of the components of the electric field, and (4) the possibility of combining MT methods with other geophysical surveys.

Vilyuis syneclise

Within the geoelectric section of the central part of the Vilyuis syneclise, a thick sequence of highly conducting Paleozoic sediments is present. These formations contribute to the left descending branch of the MT sounding curve enabling two-layer master curves to be used to determine depth to the top of the low-resistivity Paleozoic beds, and to evaluate their average longitudinal resistivity (Yakovlev and Sivitzkiy, 1969).

As may be seen from Figure 3, the Khapchagai uplift and the Linden basin are readily identified from relief on the top of the conductive formations. It is a fact that the depth to the top of the low-resistivity formations nearly coincides with the depth for seismic reflecting horizon TP, which corresponds to the base of the Mesozoic formations. As a result, in the sedimentary sequence in the central part of the Vilyuis syneclise, MT sounding can be used to separate two rock intervals and to determine their thickness and longitudinal resistivity; these two are the Mesozoic (h_1 = 2–6 km; ρ_1 = 20–40 Ω-m), and Paleogene (h_2 = 4–6 km; ρ_2 = 1–5 Ω-m) formations.

Relief of the top of the low-resistivity Paleozoic formations has been studied aerially using a telluric network (Berdichevskiy and Yakovlev, 1963). A map of the telluric field determined from magnetic variations in the period range of the descending branch of

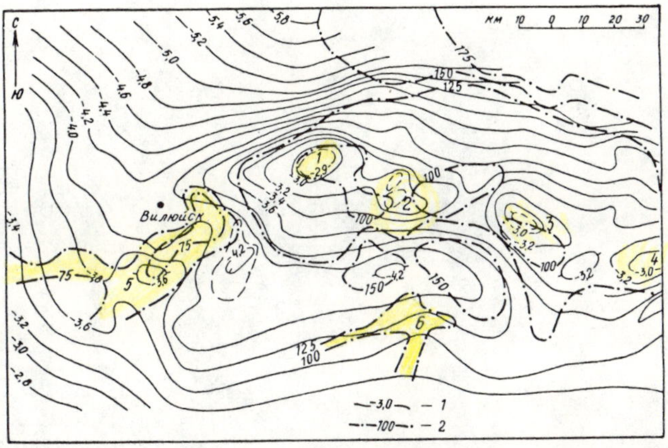

FIG. 9. Results of a telluric current survey in the area of the Khapchagai uplift (Vilyuis synclise). (1) Contour on the seismic *T-P* horizon (kilometers). (2) Contours of telluric field (arbitrary units).

the MT sounding curve (see Figure 9) was compiled. A number of local basins could be recognized on the surface of the Paleozoic formations: the mid-Vilyuis (1), the Tolon (2), the Mastakh (3), the Nedzhelin (4), the Birakan (5), and the Khaialakh (6) structural uplifts. These uplifts cause anomalously high values for the telluric intensity. The northerly Kapchagai uplift shows a progressive burial of the top of the conductive sequence toward the Linden basin (a decrease in the telluric intensity). All of these results were confirmed with subsequent seismic survey data.

Thus, data from MT surveys permitted determination of the presence of a thick conductive Paleozoic sequence in the central part of the Vilyuis syneclise, evaluation of its parameters, and determination of occurrence and regional extent of a number of uplifts. This was of great value in planning the direction of oil and gas exploration, inasmuch as the principal prospects for oil and gas are in the lower Mesozoic and Paleozoic rocks.

Northern Sakhalin

Evaluation of the sedimentary section in northern Sakhalin was made possible by the development of wide-band recording equip-

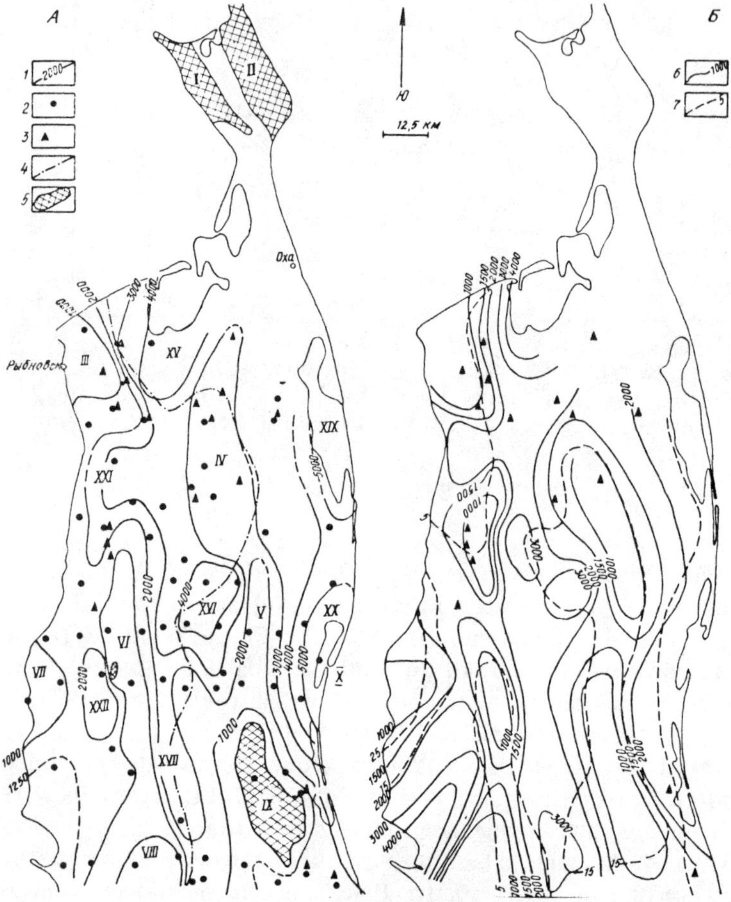

FIG. 10. Map of the elevation (A), thickness and resistivity (B) of the low-resistivity layers in North Sakhalin. (1) Contour on the surface of the conductive section (meters). (2) MT sounding location. (3) Location of well penetrating the conductive section. (4) Facies transition. (5) Outcrop of low-resistivity rocks. (6) Isopach (meters). (7) Contours of resistivity. I, II, . . ., XXII, etc., same as in Figure 4.

ment with provision for detection of short periods ($T = 0.1$ sec). This greater band width permitted satisfactory recording of the left-hand descending branch of most of the MT sounding curves.

The sandstones and shales of the low-resistivity formations of the lower to mid-Miocene section occur throughout the northern Sakhalin area. This sequence is considered at present to be potentially productive of oil and gas, so it is important to be able to study its thickness, depth, and variations in lithology.

The most satisfactory conditions for studying these beds with the MT method is found in the northwest part of Sakhalin. Here the MT sounding curves have a clearly expressed left descending branch which can be used with the appropriate two-layer master curves to determine the depth to the top of the low-resistivity horizons. In the eastern areas of north Sakhalin, the recognition of low-resistivity lower to middle Miocene beds meets with considerable difficulty because of the presence in the conducting sequence of an intermediate Dalin horizon with high resistivity and the intrusion of marine saline waters along the coastal regions.

Based on the results of interpreting the MT sounding data, a structural map was constructed on the surface of the low-resistivity horizon as well as maps showing schematic variations in its thickness and resistivity (see Figure 10). The quality of these presentations reflects the density of the survey. The most satisfactory results were obtained in the western part of Sakhalin where ρ_2/ρ_1 was less than 1/9. Comparison of MT sounding results with drilling results indicated that in this zone, the depth to the highly conducting layers as determined from MT sounding were accurate within 10 to 15 percent. Less satisfactory results were obtained in the eastern part of Sakhalin.

The maximum depth to the top of the conductive section (Figure 10a) was found at Baikal (XV), Val (XVI), and in the Piltun-Chaivin depressions (XIX-XX), where it reaches 4 to 5 km. Somewhat smaller depths (from 1.5 to 3.5 km) were found in the Chingai (XXI), the Nish (XVII), and Vagiss (XXII) depressions. Over the Landri (III) uplift, the top of the low-resistivity horizon lies at depths of several hundred meters. Over much of the Dagin (IX) uplift, the beds under consideration outcrop, but in the central area of Girgilanin (IV), and the Ossois (V) uplift, a depth varying between 1.5 and 2.5 km to this horizon is found.

The thickness of the low-resistivity sequence in northern Sakhalin (see Figure 10b) varies over a considerable range. The maximum thickness (from 4 to 6 km) is found in the Baikal, the

Nish, and the southern part of the Vagis depressions. They are unusual not only in the length of the downdropped areas, but also in the stratigraphic range represented within the sequence. In the remaining areas, the thickness of the sequence is 2 to 3 km, decreasing to 1 km or less toward the central part of the broad contemporaneous uplift.

The resistivity in the low-resistivity sequence varies within the range 5 to 25 Ω-m. These values of resistivity can be subdivided reasonably into three facies types: principally shale (5 Ω-m), sandstone and shale (5–15 Ω-m), and sandy shale (15–25 Ω-m). Relatively pure shale forms the deep conducting horizons within the Dzholokor and Lingri uplifts, and in the Chivin depression in the west and the Piltin-Chivin depression in the east. From drilling results, these sequences are essentially argillites. The shaly sand beds occur over much of the Nish-Baikal depression. On the limbs of the Dagin and Girgilanin uplifts and in the Vagis depression, siltstone formations are present. The parts of the sequence under consideration characterized by higher resistivities occur primarily in the central part of the Dagin and Girgilanin uplifts and in the east part of the Val depression.

The most interesting area from the point of view of oil and gas development is that in which shaly sand (resistivity of 5 to 15 Ω-m) occurs, as in the Nish, Baikal, and Vagis depressions, and along the flanks of the Dagin and Girgilanin uplifts. The Uzlov, Berezov, and Il'chin fields occur in this type of sequence. The deep lying, primarily shale formations that occur in the Dzholokor-Lingri uplifts and on the western edge of the Piltun-Chaivin depressions are of generally less interest in the search for oil and gas because of the poor development of sandstone reservoirs. The sandy shale sequences, having resistivities of 15 to 25 Ω-m, are favorable for oil and gas accumulations, but until now, such sequences have been virtually unexplored by deep drilling. However, their electrical characteristics indicate the presence of sandstone horizons (dominant in the sequence) and adequate thickness of shale caprock.

There is considerable practical interest in zones where transitions take place in the facies of the rocks of the sequence under consideration, because of the possible existence of stratigraphic traps. The most obvious region for exploration for oil and gas reservoirs of this type must be considered to be the transition area from Nish-Baikal depression to the Dzholokor-Langri uplift, where there is a rapid change toward the west from a sandy shale sequence to a relatively pure shale containing individual sandstone beds. The change is accompanied by a decrease in the resistivity of the sequence by a factor of 2 or 3 and an overall increase in thickness. Similar features are seen along the east edge of the Vagis depression.

Thus, MT sounding has been used to subdivide the sedimentary sequence, to recognize sections of relatively deep lying shaly sand beds distributed regionally over northern Sakhalin, and to determine their thickness, lithology, and depth in various areas. The data which were obtained make it possible to predict the potential for oil and gas saturation within these units.

The conductive beds are not always characterized on the MT sounding curves by a clearly defined descending branch. However, in such cases it is still possible to recognize them. For example, in northern west Siberia (Kopelev et al, 1976) an analysis of the type of MT sounding curve permitted outlining an area of broad development of conductive pre-Jurassic rocks. An examination of MT sounding curves processed using a relationship between apparent conductance and the effective depth of penetration of the EM field [pseudo-inversion (Yakovlev, 1975)] then made it possible to evaluate the thickness and resistivity of individual beds and to recognize a number of zones of facies transition.

It should be noted that pseudo-inversion processing for a specific horizon is most accurate when applied to subdivide a geoelectric sequence with several boundaries. For example, up to seven sections can be recognized in the sequence in the Dneprov-Donetz basin (Tomchakov and Kharchenko, 1976), and several subsections can be recognized in the Cis-Caspian basin (Faradzhev, and others, 1976) and in the Urals foothills basin (Brusyanin, and others, 1976). The presence in the sequence of an intermediate screening layer which is transparent for one of the components of the electric field makes possible obtaining independent information about the structure of two horizons. Such results have been obtained in the Moscow syneclise (see section V), the Timan-Pechor Province (Podlovilin and Gablin, 1976), the Pripyat basin (Mgu, 1976), and the Cis-Caspian basin (Brusyanin et al, 1976; Sidorov et al, 1976). Finally, it should be remembered that it is more practical to subdivide a sedimentary sequence using several geophysical techniques, each of which individually permits the evaluation of the structure in a particular horizon. An example can be provided by the results of exploration in the eastern Caucasus foothills (Konnov and Varlamov, 1971) where combined reduction of MT data, seismic refraction data, and seismic reflection data in the east Manich basin made it possible to determine the thickness of

Jurassic and Permo-Triassic beds characterized by high resistivity and to establish approximate limits on the regional distribution of these beds (Averbukh et al, 1974). Further development of capabilities for subdividing the sedimentary sequence in the future will be provided by increased accuracy in recording MT sounding curves and by provision for recording a wider range of frequencies.

IDENTIFICATION OF FAULTS FROM MAGNETOTELLURIC DATA

The recognition of faults, as one might expect, is accomplished by analyzing a variety of geologic and geophysical data which in favorable cases shows an anomaly which may be clear evidence for the presence of some degree of dislocation in the rocks. In order to understand the geologic nature of these anomalies, it is not necessary to carry out an exact analysis of their features. Under favorable conditions, sufficient information may be obtained from MT exploration.

Indicators of the presence of faults may include: (1) a high gradient for the various MT parameters; (2) a characteristic change of contours in a direction orthogonal to the general direction; (3) the presence of linear anomalies of various parameters; and (4) a sudden change in form and orientation of the polar diagrams representing the change in size of the MT parameters as a function of the direction of measurement. Such faults in the basement and the sedimentary section may sharply distort the horizontal homogeneity of the section, particularly for measurements of transverse resistivity measured along profiles crossing fault zones.

Figure 11 shows a cross-section of transverse resistivity along the profile crossing the Vilyuis syneclise in the vicinity of the Suntar uplift and the adjacent Vgyattan and Kampendai basins. In the vicinity of the sharp gradients in transverse resistivity (the contours are nearly coincident), faults are present separating the basins from the uplift. These faults are clearly expressed also in the seismic sections and the gravity map.

The locations of rapid gradient of depths on structural maps for the surface of the crystalline basement also correspond to zones of

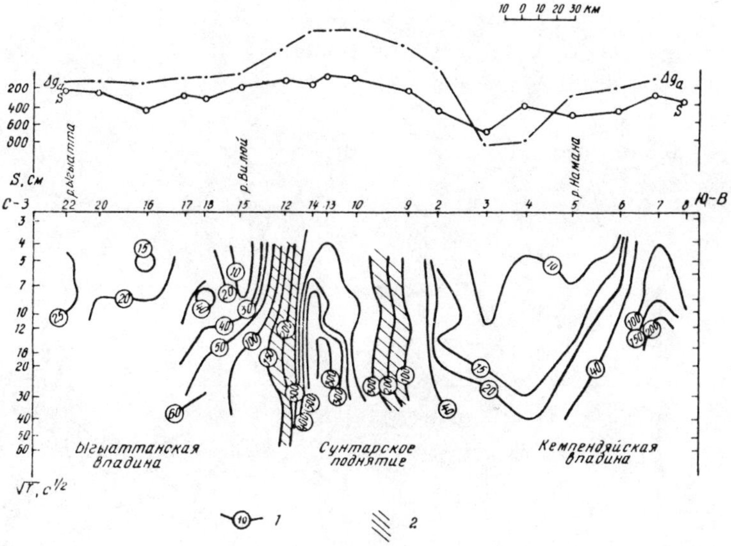

FIG. 11. Profile of $\rho_T^{\perp}$ across the Suntar uplift, and the Kempendyai and Vgiattan basins (Yakutia). (1) Contour of $\rho_T^{\perp}$. (2) Fault zone.

sharp gradient of apparent resistivity for the transverse MT sounding curves. These zones indicate the presence of faults bounding the Suntar uplift, the Kempendyai and Vgyattan basins, the Botuobin rise, and the Nyuis-Dzherbin and Berezov basins.

On MT sounding data from northern Sakhalin (Alperovich, and others, 1972), changes in the thickness of the sedimentary sequence over the various uplifts and downwarps were clearly apparent, indicating the presence of faults in the deep formations. The transverse resistivity curves indicated the presence of faults of considerable amplitude penetrating into the metamorphic basement. Moreover, on these curves it was possible qualitatively to determine the character of change in amplitude of faulting with depth. In these same regions the gradient of change of transverse resistivity with depth also showed a characteristic increase, providing evidence on the increased consolidation of the rocks in the vicinity of the faults.

The utility of the MT sounding method in studying deep tectonic features is demonstrated by the examples of west Baikal (section A-a) and east Engizpal (section B-b and C-c) as shown in Figure 12.

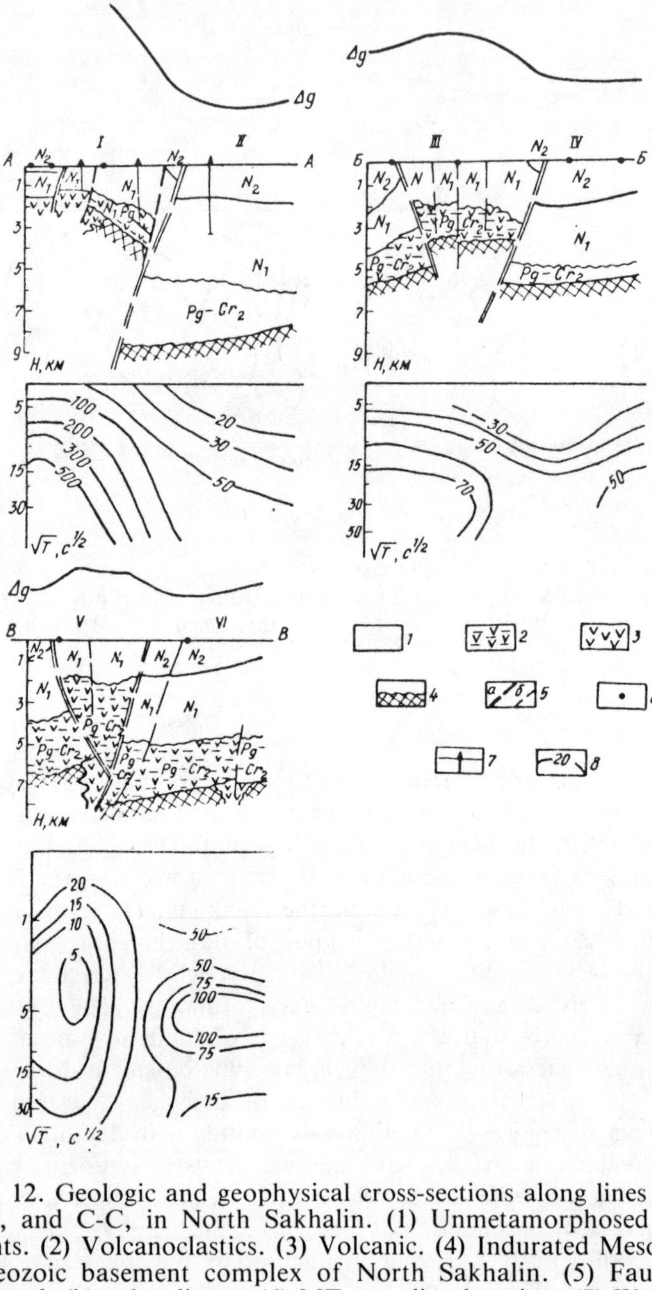

FIG. 12. Geologic and geophysical cross-sections along lines A-A, B-B, and C-C, in North Sakhalin. (1) Unmetamorphosed sediments. (2) Volcanoclastics. (3) Volcanic. (4) Indurated Mesozoic-Paleozoic basement complex of North Sakhalin. (5) Fault (a) regional; (b) subordinate. (6) MT sounding location. (7) Well. (8) Resistivity contour.

The west Baikal fault is situated on the eastern edge of Langri uplift and separates it from structures of the Baikal graben and syncline zone. From the MT data, the amplitude of faulting at the top of the metamorphic section is 4 to 5 km. On the transverse resistivity curves, over the fault zone, the gradient in change of resistivity increases with increasing period (beginning at $T = 100$ sec), which apparently is evidence for the depth of faulting and the length of time over which it has developed. Further study of the fault zone in west Baikal would be of great practical interest since it is quite possible that oil and gas fields lie hidden here.

The east Engizpal fault extends in a north-south direction along the east edge of the Engizpal and Dzholokor horst-like uplift, separating them from the Val, Nish, and Timov depressions. The fault is indicated by a gravity step and a positive gravity anomaly. The MT sounding data show a sharp change in depth to the surface of the metamorphic basement at the fault. The amplitude of offset amounts to 2.5–3 km. The east Engizpal fault, as is the case in west Baikal, has had a long history of development in the middle part of the sedimentary complex, since the sediments east and west of the fault have been deposited in different environments. The section to the west is primarily marine and shoreline shales along with shaly sand sediments (resistivity, 5 to 15 Ω-m). To the east of the fault, in the northern part of the Nish depression, the sediments are lagunal and continental, primarily sands (resistivity, 30 to 50 Ω-m). Therefore, the MT data permitted us to obtain quite meaningful information on the nature of the geologic structure in the vicinity of several major faults.

In the Moscow syneclise, two faults which have been traced in the basement provide anomalies, in particular (1) a sharp gradient in E; (2) a sharp change in the direction of the maximum axis of the absolute elipse constructed for periods corresponding to the asymptotic portion of the MT sounding curves; (3) the existence of a zone with anomalous values for the parameter N (the ratio of major to minor axis in the polar diagram for the modulus of the impedance); and (4) a sharp change in the direction of contours, and others. Similar phenomena reflecting tectonic features are also apparent from gravity and magnetic data, and from refraction data and drilling. The most important of these tectonic elements (see Figure 8) are controlled by a northwest-to-southeast graben structure, together with the southeastern slope of the Chukhlom rise and the southern slope of the Galich uplift.

In several areas such as Russia, Tadzhik, the eastern Caucasus foothills, and the Tungus syneclise, fault zones giving an anomalous MT response occur.

In the eastern Caucasus foothills, two types of such zones are recognized. In zones of the first type, the transverse MT sounding curves (relative to the long axis of the zone) are characterized by depressed values of apparent resistivity. They are particularly depressed in the vicinity of the minimums on the curves, which results in considerable distortion. The minimum of the curve is narrow and deep, and the transition from the minimum to the rising portion of the curve is very sharp. The longitudinal curves are characterized by elevated values of apparent resistivity. In zones of the second type, the longitudinal and transverse curves are lowered relative to the neighboring normal curves, and the longitudinal curve is distorted as strongly as the transverse curves.

Comparing the MT sounding curves obtained in the anomalous zones with theoretical calculations and physical models, we can develop an idea of the geologic and geophysical nature of the anomaly. In zones of the first type, the transverse resistivity curve agrees well with data from physical modeling (Kovtun et al, 1970), in particular, with results of theoretical calculations for the case of a vertical contact with *H* polarization and with practical curves obtained in the Tadzhik depression (Koshlakov, 1970). For both the field curves and theoretical curves, a rapid transition from the minimum to the rising segment was found. A comparison of models, theoretical calculations, and experimental curves suggests that the anomalous zone of the first type is observed over step-like relief on the surface of the basement which deflects the current and leads to a decrease in transverse resistivity and an increase in longitudinal resistivity.

Magnetotelluric sounding curves obtained in anomalous zones of the second type agree qualitatively with the results of theoretical calculations carried out by Kokotushkin for narrow grabens (Dmitriev and Kokotushkin, 1971). These anomalous zones may also be correlated with narrow zones of increased conductivity associated with faults in the lower sedimentary section and basement, since an increase in fracturing and porosity close to the faults affects hydrogeologic conditions and enables the upflow of high-temperature water and highly saline water. It has been established that some zones of deep faulting in the Tungus syneclise have high electrical conductivity and cause strong distortions in MT sounding curves, but there are also MT data indicating the presence of faults that are not electrically conductive. Therefore, both the types of distortions which have been described occur in the geologic structures of these regions.

A few of the anomalous zones which are defined from MT sounding data are also noted in other geologic and geophysical data. We should, however, note that there may be a lack of correspondence between the strong amplitude MT anomalies and the comparatively small amplitudes of faulting seen on seismic data.

From all of this, it follows that recognition of anomalous zones in the MT results is useful in delineating specific changes in the geologic section, which may be related to faults in the basement and sedimentary section. However, a unique interpretation of MT data is possible only in combination with data from other methods.

A very fortunate correlation of results was obtained from a variety of geophysical techniques for surveys done in the eastern Caucasus foothills. Interpretation permitted the identification of a system of regional faults and an evaluation of the role of these

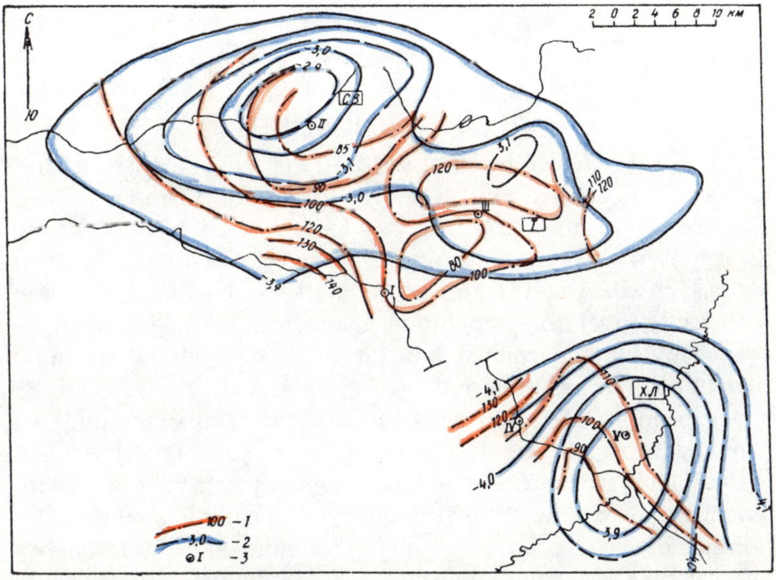

FIG. 13. Results of MT surveys of Mid-Vilyuis (CB), Tolon (T), and Khaialakh, in Yakutia. (1) Contours of telluric intensity (arbitrary units). (2) Elevation contours on the seismic *T-P* horizon (kilometers). (3) Locations of reference MT soundings.

faults in the tectonics and formation of the contemporaneous relief on the surface of the basement and boundaries in the sedimentary section. This led to several hypotheses about the character and properties of these tectonic elements. Magnetotelluric sounding data provided a basis for recognizing not only regional faults, which define the major tectonic elements, but interblock faulting as well.

These examples show that MT investigations permit recognition of zones with local geoelectric inhomogeneity, frequently associated with faulting. Future use of the MT systems for the purpose of identifying faults depends upon an increased detail of surveying, increased accuracy in calculating the various MT parameters, and the development of new methods for recognizing and interpreting recorded anomalies caused by horizontal electrical inhomogeneities.

THE UTILITY OF MAGNETOTELLURIC DATA IN RECOGNITION OF LOCAL STRUCTURES

The use of MT methods in detailed studies of geologic structures has been very limited until recently. The use of analog recording equipment has imposed a relatively low standard of precision on the results of surveys, so positive results have been obtained only in regions in which a structure of the electrical basement has been sufficiently great for recognition against the level of distortions created by inhomogeneities and the shallower beds. The telluric current method has been most widely used.

An example of the effective use of the telluric method for mapping local structures is a survey done on the scale of 1:200,000 in the Vilyuis syneclise. The telluric soundings were backed up with a network of parametric MT soundings. Depending upon the relationship between a period of micropulsations and the frequency interval for the rising segment of an MT sounding curve, it was possible to map either the top of the conductive Paleozoic section or the top of the resistant basement. The best results were obtained in mid-Vilyuis, Tolon, Khaialaka, and the Ust-Markhin areas. In the first three areas, the relief of the conducting Paleozoic beds was mapped. On a map showing the telluric field (Figure 13), three minimums in the intensity of the telluric field can be recognized. These minimums are associated with local structures recognized on the seismic *T-P* horizon. The depth to the top of the conducting sequence was determined from the descending portion of the MT sounding curves, and these depths were close to the depths of the *T-P* horizon. Consequently, the anomaly in the telluric field revealed structures in the surface of the Paleozoic sequence. Data from seismic exploration in the Khailaka area confirmed the results of the telluric survey. From the electrical data, a structure was resolved in the form of a nose extended in the north-south direction, while the reflection data indicated an ancient anacline of the metamorphic surface also with north-south extent. The results of subsequent core drilling to the Mesozoic formations and further seismic surveys confirmed the results of the telluric survey.

In the Ust-Markhin area, located on the buried northeast monocline of the Suntar Uplift, exploration was carried out to map the resistant basement surface. On a map of telluric intensity (Figure 14), positive anomalies are clearly defined which are elongate in the northeast-southwest direction. These were subsequently detailed with seismic exploration to the *K*(?) horizon (lower Paleozoic) which led to the recognition of a number of local structural highs within each of these telluric anomalies.

Thus, for conditions of the Vilyuis syneclise, MT exploration provided not only information about regional geologic structure, but also led to the discovery of local structures in the sedimentary sequence and in relief of the basement surface. In the Moscow syneclise, MT measurements also led to a solution of problems in a search for local structures. The effectiveness for solution of such problems basically is dependent on the depth to high-resistivity rocks in the second geoelectric unit (see the discussion in the first section). In zones where the strongly faulted high-resistivity rocks lie at depths less than 200 m, local anomalies of the telluric field basically correspond to relief on the surface of these rocks.

More complicated situations arise in areas where the rocks of the second geoelectric unit are buried at several tens of meters. In such cases, the anomaly in the telluric field corresponds simultaneously to the effects of relief on the high-resistivity beds and relief of the basement surface. Separation of these effects is an important problem in the application of MT exploration.

For conditions prevailing with tensional structures, which can generally be approximated with 2-D models, this problem can be

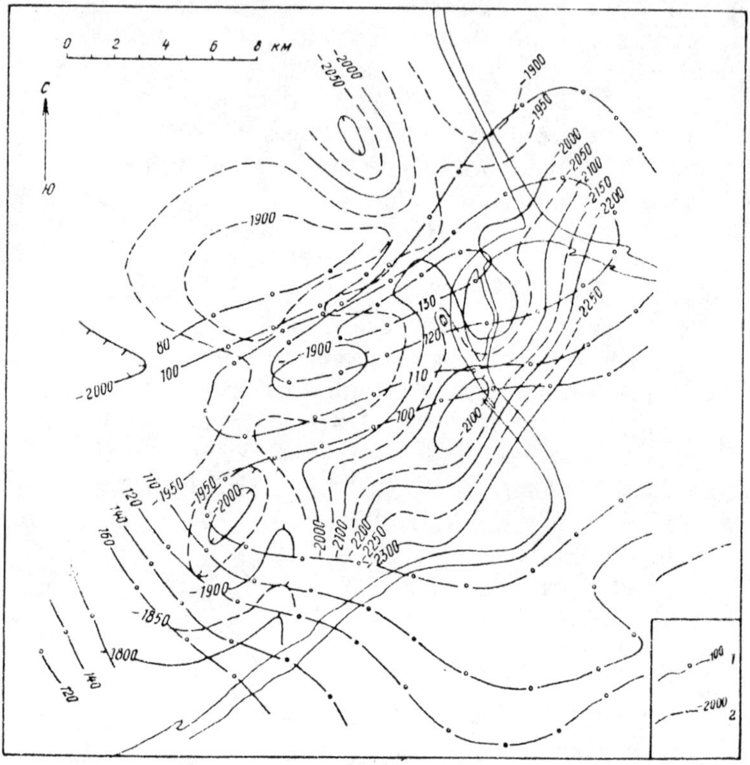

FIG. 14. Results of telluric profiling in the Ust-Markhin area (Yakutia). (1) Contours of telluric intensity (arbitrary units). (2) Depth to seismic horizon *K*(?) (meters).

solved reasonably satisfactorily. It has been shown (Berdichevskiy, 1968) that in 2-D models which contain an intermediate high resistivity screening layer, the properties of the telluric field depend primarily on its direction. The transverse field reflects primarily relief of the screening layer, while the longitudinal field is related to relief on the high-resistivity basement surface. In the Moscow syneclise, there are large extensional structures which allow the use of this approach for separating the effects created by an upper screening layer and by basement.

The problem is resolved by determining the semiaxis and the

orientation of the axis of the absolute telluric ellipses. It has been demonstrated that the major semiaxis is primarily oriented across the extent of high-resistivity Paleozoic beds, while the smaller semiaxis is oriented longitudinally to this direction. This leads to the hypothesis that the value for the maximum electric field relates basically to the transverse flow of current and consequently essentially reflects the relief on the surface of the second geoelectric unit. On the other hand, the minimum value for the electric field corresponds to the longitudinal flow of the current, which reflects relief on the surface of the basement. This hypothesis has been confirmed by correlating the values of maximum electric field and minimum electric field with depth to the first resistor and depth to the basement. This has shown that in the marginal areas of the syneclise, the magnitude of the maximum electric field indicates relief on the surface of high-resistivity beds in the upper Paleozoic; the correlation coefficient between the maximum electric field and the depth to this upper resistant unit is 0.85. By contrast, the value for minimum electric field is closely correlated with the total depth of the basement. The correlation between maximum electric field and the depth to the shallower resistant beds weakens markedly in the central part of the syneclise where the resistant upper Paleozoic beds lie at depths greater than 500 m.

The separation of local structures based on telluric profiling data has been achieved as follows. In the central part of the Moscow syneclise where the depth to the resistant upper Paleozoic carbonate sequence is no greater than 200 m, we use a map of the average electric field, which reflects the relief on the surface of the high-resistivity rocks. Where the depth of burial for these rocks is 500 to 600 m, we can use a map of a maximum electric field which reflects relief on the surface of the basement. Along the edges and near the edges of the syneclise, the relief on the high-resistivity beds is studied using maximum electric fields, while relief on the basement is studied from maps of minimum electric field.

Figure 15 shows a map of the maximum electric field. On this map it is clearly evident that most of the structures identified with seismic reflection surveys correspond to high values for maximum electric field. This is noted also in the case of large-scale structures (the Dyakanov uplift) as well as smaller structures (Yakovlev, north Neis, Savin, and Privolsh rises).

In areas where there are little or no seismic reflection data, one may expect the telluric method to find elevations in the surface of the upper Paleozoic complex.

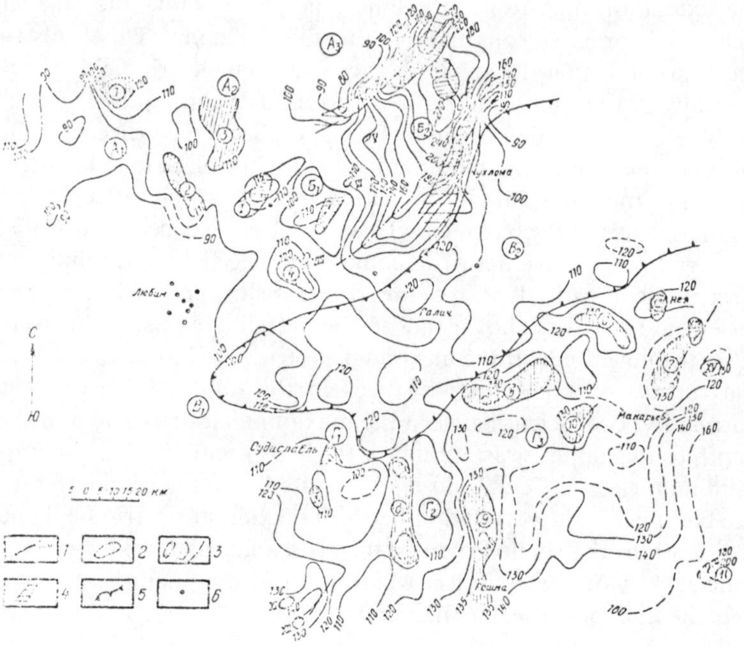

FIG. 15. Contour map of the parameter E_{max} and areas thought to be structurally high in the upper Paleozoic complex (from A.L. Sheinkman). Tectonic elements in the upper Paleozoic complex (from V.N. Gordasnikov): A-Gryazevetz-Tarnog system of basins (Al-Gryazevetz depression; A2-Kamen graben; A3-Soligalich basin); B-Ribino Sukhon system of structural highs (B1-Talitz depression; B2-Chukhlom ridge); C-Yaroslav-Nikol system of structural highs (C1-Yaroslav depression, C2-Galich uplift); D-south edge of Moscow Syneclise (D1-Sudislav graben, D2-Kineshmin depression, D3-Kadis uplift zone). (1) Contours of the parameter E_{max}. (2) Areas of interpreted local structures based on electrical prospecting data; 1-Dikarev, 2-Ramen, 3-Vokhtog, 4-Buis, 5-Filin, 6-Losik, 7-Makarev-Unzhin, 8-Sudislav-Kineshmin, 9-Reshmin, 10-Kadim, 11-Naumov. (3) Locations of structural highs and structural axes as determined from seismic reflection surveys. I-Dyakonov, II-Soligalich, III-Burdakov, IV-Bezimai, V-Yakovlev, VI-Antonov, VII-Ivankin, VIII-Buis, IX-Sudsilav, X-Savin, XI-Provolzh, XII-Pless, XIII-North Neis, XIV-Kadin, XV-Unzhin. (4) Flexures in the upper Paleozoic complex basic on seismic reflection data and drilling. (5) Outline of areas where the depth to the top of the upper Paleozoic is no more than 550 m. (6) Deep wells.

In the northwestern part of the map showing maximum electric field, within the boundaries of the Gryazevetz-Tarnog system of basins (A), we can predict the presence of local structures in the Dikarev Raman and Vokhtog areas.

In the southern part of the Ribin-Sukhon system of structures (B), the Buis area has been recognized. In the southern Galich basin (B_2), the Filin zone has been recognized, on the eastern boundary of which seismic reflection data have located structures in the Devonian section. In this same region, we would predict the existence of the Losin and Makarev-Unzhin zones which extend in the north-south direction. Along the northern limb of these zones, seismic reflection work has located an uplift (Neis) in Permian beds, while to the south there are structures in the Devonian. Along the south edge of the syneclise, two broad zones were recognized (the Sudislav-Kipeshmin and Reshmin) which had their long dimensions in the north-south direction. Within these zones, anomalies in the maximum electric field with modest amplitude are noted. In these same regions of the syneclise, the Naumov and Kadiis areas have been predicted, in addition, in the central part of the latter, seismic reflection profiling has located an axial zone of uplifted layers of Permian rock.

A map for the behavior of the minimum electric field was analyzed in a similar manner. A number of zones of elevation on the electrical basement surface are shown in Figure 6; for some conditions these correspond to the surface of the crystalline basement, and for others, possibly to horizons in the lower part of the sedimentary section.

In the eastern part along the southern border zone of the syneclise, we predicted the existence of the Naumov local structural uplift. A number of uplifts were predicted in the Reshmin and Parfenev basement uplifts, these being at Reshmin, Filin, Neis, Parfenev, and Losin. In the zone immediately adjacent to the west of the Reshmin uplift, we predicted the presence of local structures in the Sysanin, Sudislev, and Prevolsh areas. In the Galich basin, we have predicted the existence of the Buis, Galich, Burmakin, Orekhov, and Chukhlom structural highs. In the area of the Prechistin graben, we have identified the Ramen and Vosin areas with local structures. Adjacent to this zone to the south is the large-scale Dyakonov uplift. Subsequent seismic exploration has confirmed the existence of the Dyakonov, Buis, and Orekhov structural highs.

Simultaneous analysis of maps of both maximum and minimum

electric fields leads to results on the relationships between structural uplifts predicted on the surface of the high-resistivity Paleozoic beds and those on the basement surface. Most of the structural highs found in the basement were also found directly above in the upper horizons of the sedimentary sequence. It is true that in some cases, such as at Dyakonov, there is nearly perpendicular orientation of the structural axis for the basement and for the Permian beds, while in other cases, as at Buis, Neis, and Naumov, some shift in the position of the structural highs is found. Structural relief on the basement surface (as at Galich, Burmakin, Susanin, and Parfenev) is not reflected in the upper part of the sedimentary section. Some of these local structural highs have been recommended for high-priority verification by detailed seismic exploration.

Thus, analysis of telluric data has permitted the identification of local structures in the sedimentary complex and on the basement surface in the Moscow syneclise, as well as their relative structural positions.

The use of the telluric method has proved to be highly effective for studying salt tectonics. Examples of such applications are investigations carried out at Pripyat and in the Cis-Caspian basins. The results of these studies indicate that with the aid of MT profiling, the search for and delineation of local salt structures permits detection not only of major salt masses, but also of individual salt domes and intrusives, and permits study of other details of their structure.

In the case of high transverse resistivity in an intermediate resistance screening layer, the practicality of using the telluric method (as might be expected) is limited for studying beds beneath such screens. Even in a medium which may be considered 2-D, the use of the longitudinal component of the electric field (E-polarization) does not always permit the study of local structures in the sequence beneath the screening layer. In this respect, the MT method has essential advantages. As is well known, a thin screening layer has no effect on the magnetic field, and inhomogeneities in the sequence beneath the screening layer can be studied with reasonable resolution.

During the period from 1966 to 1968, practical results were obtained in the Tadzhik depression which indicated that the study of the MT field permitted mapping of local structures in the subsalt sequence. This section is considered to be upper Jurassic in age and has good prospects in the search for oil and gas fields. Because it is assumed that the structures in the lower part of the sedimentary section are inherited from the surface of the basement, the question reduces to the recognition of basins and uplifts (possibly fault blocks) in the basement.

Favorable geologic factors for large-scale regional MT surveys include linear tectonic patterns, permitting the examination of horizontal inhomogeneities as 2-D in form and high-amplitude local subsalt structures (up to 1/3 of the depth of burial). Surveys were carried out using MT profiling on a network of 1 by 2 km with a range of periods from 15 to 150 sec. Magnetotelluric sounding stations were located on a coarser network (5 by 10 km) with a period range from 10 to 1200 sec. Profiling was done based on a single base station with a radius of 30 to 40 km. In the later parts of the survey, synchronized detection of the horizontal components of the magnetic field H_x and H_y was carried out. At the present time, the vertical component H_z is being recorded to provide supplementary information. The measuring rays were oriented along and transverse to the basic tectonic lineaments.

The parameters ν_{xx} and ν_{yy} were used as the basic characteristics of the field, these being the coefficients of linear correlation:

$$H_x^q = \nu_{xx} H_x^p + \nu_{xv} H_y^p ,$$

and

$$H_y^q = \nu_{yx} H_x^p + \nu_{yy} H_y^p ,$$

where the indices q and p indicate the field and base stations, respectively, and H_x and H_y are the components of the magnetic field along the measurement lines. Both the principal impedance and the skew impedances were determined at the MT sounding reference points. Subsequently, the parameters ν_{xy} and ν_{yx} were used as a measure of the horizontal inhomogeneity.

Values for all of the quantities which have been listed were determined by graphical methods which are taken to be generalized and highly developed modifications of the apparent impedance method. The results indicate that the method used in reducing the data has given very accurate results in comparison with other methods (such as least squares, vector polarization, etc.) under conditions of marked lateral inhomogeneities in the section and quasi-linear polarization of the field.

As an example of the particular use of the MT method in the search for local uplifts and subsalt section, we can consider the results of surveys in the area of Dosmak-Agi-Aktai, located in the western part of the depression (the Surkhan-Darin zone, Uzbekistan). These results are shown on Figure 16.

The parameter ν_{xx}, corresponding to E-polarization, is characterized by well-defined zones of minimum values (80 arbitrary units) and maximum values (115–120 arbitrary units); these zones correspond to uplifts and basins on the reflecting seismic horizon

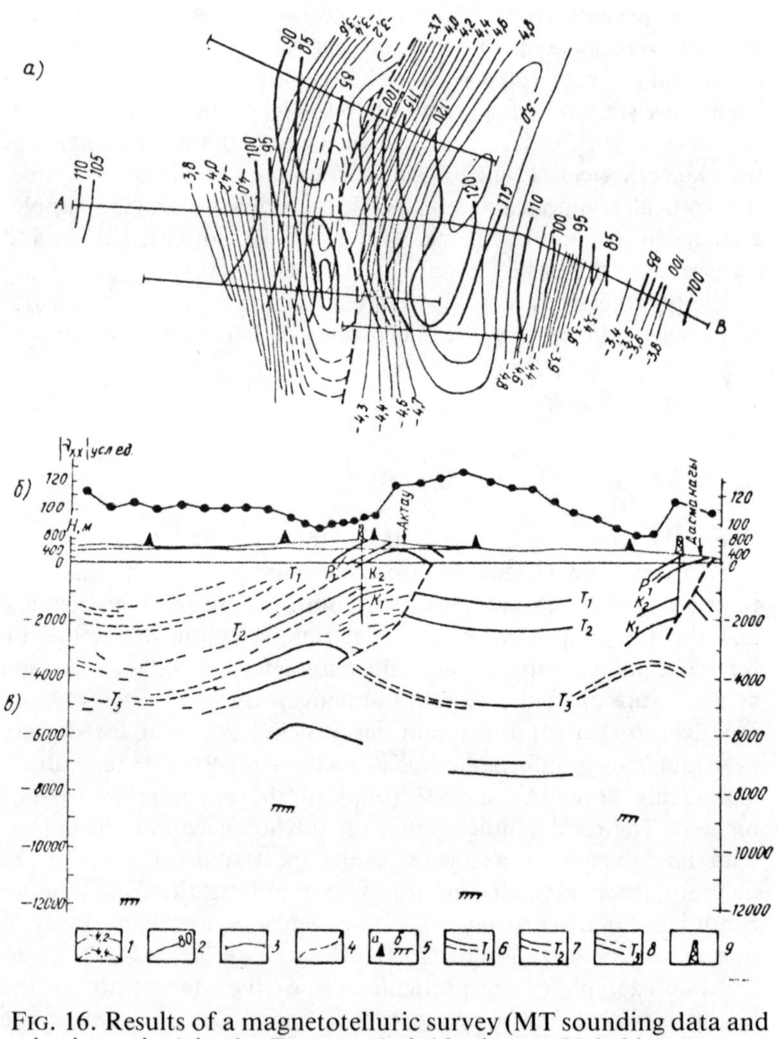

FIG. 16. Results of a magnetotelluric survey (MT sounding data and seismic section) in the Dasman-Agi-Aktai area (Uzbekistan).

T_3, stratigraphically identified with the subsalt Jurassic formations. The magnitudes of the anomalies are no less than 20 percent of the average level of the field. The amplitudes of the anticlinal structures, as determined by seismic reflection, are of the order 25 percent. The configurations of the anomalies in measured maximums and minimums practically coincide with the detailed structural map on the T_3 horizon. It should be noted that the parameter ν_{yy} corresponding to H-polarization is about 100 arbitrary units (with a precision of 5 percent) at all points on the profile. This confirms to some degree the high degree of significance of the 2-D model to the real medium, and on the other hand the accuracy of the orientation of the measurement array with the axis of the electrical inhomogeneities and across them. At the same time, the graph of ν_{xx} does not reflect the structure in the upper part of the section which was clearly mapped using the telluric current method (H-polarization). In addition to the relation of the anomaly in the parameter ν_{xx} as a consequence of structure in the subsalt formation, we should indicate a poor correlation between the values of ν_{xx} and the total longitudinal conductance of the Neogene-Paleogene formations, as obtained from Schlumberger sounding. These values reflect an electrical basement as being the surface of carbonate formations of the lower Paleogene (Bukhar formation).

A very important aspect in evaluating the effectiveness of methods based on EM fields is the degree of resolution of the anomalies in relation to the frequencies of the variations which are measured. As indicated by E-polarization calculations for models of uplifts in the surface of a reference horizon with high-resistivity layers on top, a change in relative intensity of the magnetic field is detectable if the wavelength λ is greater than half the width of the uplift. For longer waves, close to the dimension of the inhomogeneity, the width of the anomaly corresponds to the width of the structure. Under conditions which are realistic, i.e., for a resistivity of 5 to 10 Ω-m even with a period of 10 sec, we have wavelengths of 10 to 20 kilometers. This size coincides with the transverse dimension of the assumed local uplifts.

The minimum relative intensity over an uplift is sharper and clearer, when the ratio λ/h_o is greater, where h_o is depth to the electrical basement over the uplift. As λ/h_o increases to 5 to 20, a relative intensity of the magentic field varies by only 5 percent (see Figure 17b). Furthermore, with use of a full decade of periods (T = 15–150 sec), the wavelength for constant resistivity changes only by a factor of 3. Therefore, it can be verified that the error caused by

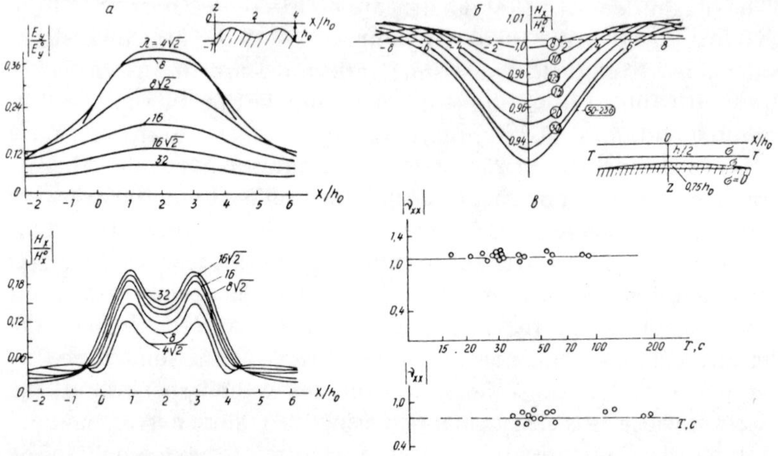

FIG. 17. Relationship of the intensity of the MT field to wavelength (E-polarization). (a) Model with local inhomogeneities (after Drobnitz, 1973); (b) uplift of the basement surface covered by a thin layer (after Berdichevskiy et al, 1966); (c) observed relationship beteween ν_{xx} and period T.

change in wavelength on increasing the bandwidth is small and that over the entire range of short-period micropulsations (kpk), the anomaly in the intensity of the magnetic field caused by local inhomogeneities in the electrical basement will be satisfactorily recognized.

Figure 17 shows the indicated practical relationship of the parameter ν_{xx} to the period of oscillation at points over the crest of an uplift of the electrical basement and over its flanks. They are well represented by a straight line parallel to the period axis, which confirms the essential independence of the behavior of the horizontal magnetic component (H_x) with respect to the frequency of the observed field.

Figure 17 shows theoretical curves for the field components computed for models containing local inhomogeneities. As may be seen, the most diagnostic profile is provided by the magnetic component H_x (E-polarization).

Actual field results similar to the example considered above have been obtained in other areas of the Tadzhik depression (Khodzha-Kazian, northern Aruktai, and others). At the present time, detailed seismic reflection surveys at a scale of 1:50,000 are being carried out as part of an integrated geophysical program to locate local structures in the subsalt deposits in this structurally complicated area.

Despite the low productivity of the method, the use of MT sounding in similar applications for detailed surveys appears to be quite feasible where electrical basement is at great depths. Evidence for this is provided for MT soundings in the Gornostiev area (Kerchen Peninsula). Here at the base of the Mesozoic formations, which is considered to be an electrical horizon at a depth of 5 to 8 km, a local broad amplitude structure was outlined, which was subsequently confirmed by drilling and seismic work. Interesting results were also obtained using detailed MT sounding surveys in Azerbaizhan.

With the introduction into commercial operations of digital electrical recording equipment, including a digital automatic data reduction system, the productivity of MT sounding can be markedly increased (by a factor of 5 to 7) as well as its accuracy. These two developments open broad applications for the use of MT sounding for detailed surveys. One of the first results has been obtained in the Dyakonov area of the Moscow syneclise.

Studies of the local structure on the surface of the basement in this area are of great value because the relief of this surface is directly reflected in the oil and gas prospects in the pre-Devonian formations. This also provides a basis for carrying out a detailed MT survey in this area at a scale of 1:100,000 using the digital electrical prospecting system. Interpretations were based upon a relationship between cumulative conductance S_T and the interval depth of investigation h_T. The values for S_T and h_T were determined either from values of average impedance or from values of the principal impedance, corresponding to the major axis of the polar diagrams constructed for periods of 80 to 100 sec. This interval of periods corresponds to the rising branch of the apparent resistivity curve.

The interpretation was completed by correlation of the inflection points on the curves under interpretation, determination of the coordinates of these points, and the subsequent computation of thickness, longitudinal conductance, and resistivity of the selected layers. From the interpretation of 125 MT sounding curves, a structural map on the surface of the basement in the Dyakanov area was constructed, as shown in Figure 18. This structural map shows the Dyakanov uplift which has a nearly north-south strike and which is bounded to the northeast by faults. This structural high is an uplifted block of the basement, with the top lying at a depth of 3.2 to 3.4 km as indicated in wells D-1. The faults which bound the structural high are mutually orthogonal and extend in the southeast or southwest directions. Furthermore, the more northerly well, D-2, penetrated one of these fractures, giving credibility to the results of the electrical survey. On crossing the central part of the syneclise in a northeasterly direction, the northern boundary of the Soligalich-Roslyatin graben is clearly seen. To the north of the faults, there is a relatively rapid increase in the thickness of the Ordovician section, together with the appearance in the section of reef rocks (thickness of about 500 m) which are absent on the crest of the Dyakanov structural high. The amplitude in the northeasterly direction of the Dyakanov structure is about 600 m, and in the northwesterly direction, about 300 m. The south flank of the structure is relatively flat. In the northwestern part of the region studied, the fault-bounded Raman structural high was noted, striking northwest to north. The minimum depth to the crest of the Raman structure is about 3000 m. The northern limit of the structure was not traced.

The south Demyanov structure is located farther north of the Dyakanov uplift and is indicated on the basis of two points. This structure can be recognized only because on the map it corresponds well with the anomaly detected with a high precision gravity survey.

Within the limits of these recognized uplifts, a change in the average resistivity of the sedimentary section can be seen, which provides evidence for the change in character of the geoelectric section. Low values of average resistivity (3–4 Ω-m) were found in the southern limb of the Dyakanov structure, while over the rest, even lower values (2–3 Ω-m) in common with the values north of the regional faults. It provides a basis for assuming that the observations in this part of the area reflect an increase in the thickness of the third terrigenous low-resistivity complex.

The accuracy of geologic interpretation may be judged on the brief results of interpreting conductance values obtained from well logs in wells Dyakanov 1 and 3, where an error in determining depth to basement from magnetotelluric sounding did not exceed 5 percent. The results of this detailed regional MT sounding survey for the Dyakanov area made it possible to obtain new data on the structural relief of the crystallized basement, as well as the accurate

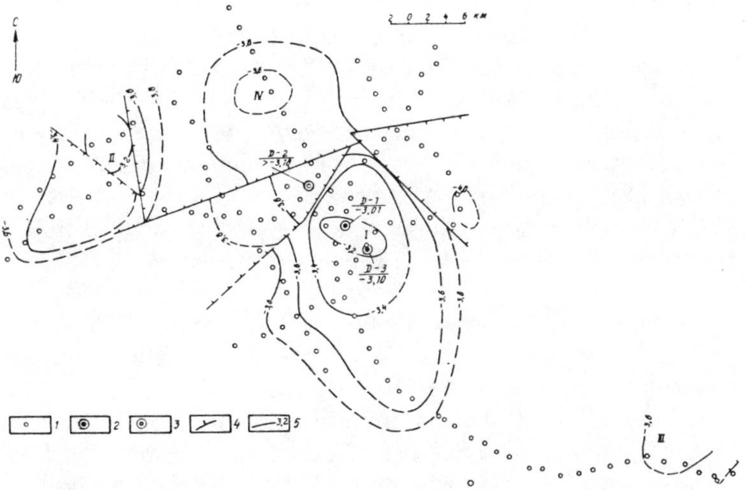

FIG. 18. Map of basement relief in the Dyakonov area, from MT data. Moscow Synclise (from A. L. Sheinkman). (1) MT sounding locations and depths to basement. (2) Wells serving as control points for depth to basement. (3) Deep wells that do not penetrate basement. (4) Faults. (5) Contours on the basement surface (kilometers). Basement highs — I, Dyakonov; II, Ramen; III, Orekhov; IV, Demyanov.

location of local strong structure highs and the recognition of the number of other uplift and disturbed blocks of the basement.

There may be opportunity for improvement of MT surveying through the use of several MT stations operating synchronously to record simultaneously variations in the MT field over a wide range of frequencies. The results would include, in addition to the impedance characteristic at each point, a space-frequency distribution of the electric and magnetic components of the EM field which would permit a more complete geologic interpretation.

A very detailed survey done using the digital recording system in the Dneprov-Donetz basin permitted the study of the morphology, location, and extent of salt domes. The character of the anomalies in the electric and magnetic fields as plotted on frequency pseudosections completely defined the structure of the domes, which had been originally established on the basis of geologic data and the results of quantitative MT sounding interpretation.

As already noted, the fundamental advantages of digital electrical prospecting recording equipment, particularly the high accuracy of recording the MT field and of reducing the data, have increased productivity to permit the use of MT sounding for detailed investigations in the solution of a number of precise problems. Even the first results from these surveys have confirmed in principle the results of work carried out earlier on the existence of a significant number of local anomalies of various characters in the MT field. The recognition of these anomalies and their identification in terms of some type of geoelectric inhomogeneity is one of the most challenging problems in MT investigations.

Thus, in White Russia, western Siberia, and the Timan-Bechor Province, as well as a number of other regions, some significant anomalies can be assumed to reflect the influence of oil and gas reservoirs. There is reason to believe that MT surveys may be utilized also for studying the general nature of the basement.

CONCLUSIONS

This report on the nature of results of MT surveys in a number of regions in the U.S.S.R. has shown that the integration of various modifications of MT exploration permits in a wide variety of geologic problems the solution of a broad range of geologic questions, such as mapping a relief on the surface of the basement or some other high-resistivity formation, the subdivision of the sedimentary section, tracing of zones of a lithological complex, and the identification of faults.

The possibility of resolving these questions is in general determined by the geoelectric conditions, including among other things the nature of the variation in resistivity with depth and the lateral homogeneity of the physical properties of the rocks.

The most significant results have been obtained in regional investigations, and up to recently this has been the main application of such exploration. The use of MT sounding has permitted obtaining useful data on the thickness of the sedimentary sequence. The results of MT sounding both in combination with results of other geophysical methods or by itself has made it possible to construct structure maps on the surface of high-resistivity formations in many areas.

Within any region, data from MT prospecting have revealed major faults in the basement and sedimentary section and led to hypotheses on the characteristic structures of these zones.

No less important results have been obtained in subdividing the sedimentary section. In a number of areas, MT sounding has permitted not only establishing the existence of thick conducting formations in the section but also noting boundaries and their dip, evaluation of thickness and resistivity of these formations, and determination of depth of burial and facies-lithological changes in character.

For cases where intermediate screening layers exist in the section, conditions are determined for application to problems related to structure on the surface of high-resistivity formations and on the surface of the screening layers. This has led to results on the relationship of structure on the basement and in upper members of the sedimentary section.

In individual cases with favorable conditions, it has been possible to identify local structures either on the surface of the basement or at the top of intermediate high-resistivity formations, and to study the detailed structure on the surface of the basement for specific structures.

In metamorphic areas with linear tectonic control and a thick conducting sequence, it has been established that high-sensitivity short period variations in the horizontal magnetic field are related to local inhomogeneities in the top of the basement. In this respect, high-resistivity evaporite-carbonate formations which may be present in the sedimentary section present no barrier to the study of lower horizons.

This monograph indicates that the most effective method of MT exploration is MT sounding. Other methods, including magnetotelluric profiling and telluric current surveys, are less informative, but they have the advantage over MT sounding that they may be carried out more rapidly. It is obvious that the prospects for future use of MT methods depend upon an increase in their information content and reduction in their time requirements. The wider commercial use of digital recording equipment developed by the National Institute for Petroleum Geophysics in field investigations, together with reduction of data with a computer, would simultaneously solve both of these problems.

Glossary[1]

Allochthon
A mass of rocks which has been moved from its original site of origin by tectonic forces, as in a thrust sheet or nappe.

Anteclise (or anticlise)
A positive or uplifted structure of the continental platform; it is broad in regional extent (tens to hundreds of thousands of square kilometers) and is produced by slow crustal upwarp during the course of several geologic periods.

Syneclise
A negative or depressed structure of the continental platform; it is of broad regional extent (tens to hundreds of thousands of square kilometers) and is produced by slow crustal downwarp during the course of several geologic periods.

Anticlinorium
A composite synclinal structure of regional extent composed of lesser folds.

Synclinorium
A composite synclinal structure of regional extent composed of lesser folds.

Anacline
A valley that descends in a direction opposite to that of the general dip of the underlying strata it traverses.

Aulacogen
A fault-bounded intracratonic trough or graben.

Neogene
Interval of time incorporating the Miocene and Pliocene of the Tertiary period; the upper Tertiary. When the Tertiary is designated as an era, then the Neogene together with the Paleoene may be considered to be its two periods.

Paleogene
Interval of geologic time incorporating the Oligocene, Eocene, and Paleocene of the Tertiary; the lower Tertiary.

Definitions of geologic terms are from the "Glossary of Geology", A.G.I., 1972, quoted by permission.

Longitudinal conductance
Conductance in a vertical direction through the combined layers of the geoelectric section.

Longitudinal resistivity
Apparent resistivity of the geoelectric section measured from the surface using electrodes spaced parallel to strike. Also termed transverse magnetic, TM, or ρ_{-a}.

Transverse resistivity
Apparent resistivity of the geoelectric section measured from the surface using electrodes spaced normal to strike. Also termed transverse electric, TE, or $\rho_{||a}$.

Type A
A resistivity model of three layers, numbered from the top, having the relationship

$$\rho_1 < \rho_2 < \rho_3 .$$

Type H
Resistivity model of three layers, having the relationship

$$\rho_1 > \rho_2 < \rho_3 .$$

Type K
Resistivity model of three layers, having the relationship

$$\rho_1 < \rho_2 > \rho_3 .$$

Type HA
Resistivity model of four layers, having the relationship

$$\rho_1 > \rho_2 < \rho_3 < \rho_4 .$$

Type KA
Resistivity model of four layers, having the relationship

$$\rho_1 < \rho_2 > \rho_3 < \rho_4 .$$

Type KHA
Resistivity model of five layers, having the relationship

$$\rho_1 < \rho_2 > \rho_3 < \rho_4 < \rho_5 .$$

REFERENCES

Abdullaev, A. A., Gadzhiev, A. N., and Novruzov, E. S., 1974, Results of electrical exploration by regional profiling over the Kurin basin: Neft. geol. i geofiz., no. 5.

Al'perovich, I. M., Bubnov, V. P., Varlamov, D. A., Konnov, Yu. K., Konovalov, Yu. F., Koshlakov, G. V., Chernyavskiy, G. A., Sheinkman, A. L., and Yakovlev, I. A., 1977, Effectiveness of magnetotelluric exploration for studying regions with oil and gas prospects in the USSR: Moscow, OTzN-TI.

Al'perovich, I. M., and Chernyavskiy, G. A., 1973, Thickness of sedimentary rocks in North Sakhalin from MTS data: Geol. nefti i gaza, no. 6

Al'perovich, I. M., Kharakhinov, V. V., and Chernyavskiy, G. A., 1972, Study of faulting in North Sakhalin from MTS data: Geol. nefti i gaza, no. 4.

Al'perovich, I. M., Kononov, V. E., and Chernyavskiy, G. A., 1973, General facies characteristics of the lower mid-Miocene beds of North Sakhalin (from MTS data): Nefti. geol. i geofiz., no. 5.

American Geological Institute, 1972, Glossary of geology.

Anishchenko, G. N., 1976, Deep magnetotelluric surveys in the central part of the Russian platform, *in* Geoelectric and geothermal studies: A. Adam, Ed., Budapest, Akad. Kiado, p. 628-633.

Anishchenko, G. N., Mitrofanov, K. P., and Romanenko, V. P., 1975, Telluric mapping and its role in integrated geophysical surveys for studying salt tectonics in the Cis-Caspian basin: Prikla. Geofiz., no. 77.

Avagimov, A. A., Al'perovich, I. M., Varlamov, D. A., Dubrovskiy, V. G., Konnov, Yu. K., and Chernyavskiy, G. A., 1976, no title, Fizika Zemli, no. 9.

Avaginov, A. A., Ashirov, T., Dubrovsky, V. G., and Nepesov, K., 1976, Deep magnetotelluric surveys in Turkmenia and Azerbaijan, *in* Geoelectric and geothermal studies: A. Adam, Ed., Budapest, Akad. Kiado, p. 652-665.

Averbukh, A. G., Briotman, A. R., Bulgakova, I. A., Varlamov, D. A., Geisherik, G. M., Goev, V. I., Grebneva, I. L., Epifanov, A. A., and Konnov, Yu. K., 1974, Regional geological structure of the Eastern Caucasus Foothills from integrated geophysical surveys: Proc., series 9, VIEMS.

Averbukh, A. G., Broitman, A. R., Varlamov, D. A., Grebneva, I. L., and Konnov, Yu. K., 1973, A method of regional geophysical investigation of the oil and gas potential of a region: Topical scientific collection, Neft. geol. i geofiz., VNIIOENG.

Baranskiy, L. N., 1970, Spatial distribution of the amplitudes of Pi2 geomagnetic pulsations: Fizeka Zomli, v. 12, p. 69-75.

Berdichevskiy, M. N., 1968, Electrical exploration by the method of magnetotelluric profiling: Moscow, Nedra monograph.

Berdichevskiy, M. N., and Dmitriev, V. I., 1976, Basic principles of interpretation of magnetotelluric sounding curves, *in* Geoelectric and geothermal studies: A. Adam, Ed., Budapest, Akad. Kiado, p. 165-221.

Berdichevskiy, M. N., Dmitriev, V. I., Yakovlev, I. A., Bubnov, V. P., Konnov, Yu. K., and Varlamov, D. A., 1973, Magnetotelluric sounding in horizontally inhomogeneous media: Fizika Zemli, no. 1.

Berdichevskiy, M. N., Obukhov, G. G., and Rabinovich, B. I., 1970, Prikla. Geofiz., no. 60.

Berdichevskiy, M. N., Savadskaya, T. N., and Chernyavskiy, G. A., 1970, Magnetotelluric sounding in tensional settings: Fizika Zemli, v. 12, p. 76-80.

Berdichevskiy, M. N., and Yakovlev, I. A., 1963, Advances in the telluric current method: Razvedka i okhrana nedr, no. 3.

Bezruk, I. A., and Lakhtionov, V. O., 1977, Evaluation of the adequacy of determining impedance in the reduction of magnetotelluric variations: Prit. Geofiz., no. 89, p. 80-87.

Briotman, A. R., Grebneva, I. L., Geisherik, G. M., Epifanov, A. A., Konnov, Yu. K., and Varlamov, D. A., 1971, On evidence for tectonic disruptions from integrated geophysical studies in the eastern Caucasus foothills: Neft. geol. i geofiz., no. 1.

Brunelli, B. E., Dobrovolskiy, M. A., Kootun, A. A., and Kuznetsov, N. S., 1969, Some results of magnetotelluric profiling and sounding modelling of inhomogeneous structures, *in* Magnitotelluricheskie metodi izuchenuja stroeniya zemnoi kori i verkhnei mantii: Moscow, Nauka Publ. House.

Brus'yanin, L. A., Gur'ev, Yu. K., and Lyapustin, B. A., 1976, Geological effectiveness of magnetotelluric methods in Orenburg state, *in* Electromagnetrie zondirovania: Moscow, Moscow State Univ. Press.

Chernyavskiy, G. A., 1966, Concerning questions about the method of apparent impedance: Razv. Geofiz., no. 13.

Chetayev, D. N., 1970, Field structure of short period geomagnetic variations and magnetotelluric sounding: Fizika Zemli, v. 2, p. 52-56.

Dmitriev, V. I., and Kokotushkin, G. A., 1971, Collection of reference curves for magnetotelluric sounding in inhomogeneous media: Moscow, Moscow State Univ. Press.

Dmitriev, V. I., Berdichevskiy, M. N., and Kokotushkin, G. A., 1975, Collection of reference curves for magnetotelluric sounding in inhomogeneous media, part IV: Moscow, Moscow State Univ. Press.

Drobnitza, V. V., 1972, Plane wave electromagnetic field in media with curvilinear boundaries: Novosibersk, Acad. Sci. Publ. House, Siberian Div.

Dubrovskiy, V. G., 1976, Results of magnetotelluric investigations in Turkmenia, *in* Electromagnetnie zondirovaniya, ch. 2: Moscow, Moscow State Univ. Press.

Faradshev, A. S., Gambarli, R. E., Kakhramanov, K. K., and Manafli, A. I., 1976, Magnetotelluric sounding in Axerbaidjan, *in* Elektromagnetnie zondirovanie, v. 2: Moscow, Moscow National Univ. Press.

Feldman, I. S., 1976, On the nature of conductive layers in the earth's crust and upper mantle, *in* Geoelectric and geothermal studies: A. Adam, Ed., Budapest, Akad. Kiado, p. 721-730.

Gordasnikov, V. M., and Triotzkiy, V. N., 1966, Mid-Russian allochtogen — Ancient structure of the Moscow syneclise: Sooetskaya Geol., no. 11.

Gugunava, G. E., Dadunashvili, N. S., Kebuladze, V. V., Lashkhi, A. S., Tatishvili, O. V., and Tavartkiladze, Sh. A., 1976, Magnetotelluric sounding in Georgia, *in* Elektromagnetnie zondirovaniya, v. 2: Moscow, Moscow National Univ. Press.

Kaufman, A. A., and Tabarovskiy, L. A., 1970, Basic theory of magnetotelluric sounding in media with gentle structures: Novosibersk, Nauka Publ. House, Siberian Div.

Kalenov, E. N., 1970, Geological effectiveness of oil-and-gas electrical prospecting: Moscow, Nedra Publ. House.

Kalenov, E. N., 1970, Geological use of the results of magnetotelluric exploration: Moscow, Nedra Publ. House.

Konnov, Yu. K., and Varlamov, D. A., 1971, Concerning some applications of electrical prospecting with the magnetotelluric method for studying the basement in the Eastern Caucasus foothills: Neft. geol. i geofiz., no. 8.

Kootun, A. A., and Chichirina, N. D., 1976, Deep magnetotelluric surveys in the northwestern part of the Russian platform, *in* Geoelectric and geothermal studies: A. Adam, Ed., Budapest, Akad. Kiado, p. 620-627.

Kootun, A. A., Gladkiy, T. D., Dobrovolskiy, M. A., and Nepesov, K., 1970, Modeling the magnetotelluric field over a basement uplift: Fizika Zemli, no. 12.

Kopelev, Yu. S., Biserov, A. V., and Sisoev, B. K., 1976, Magnetotelluric investigations in West Siberia, *in* Electromagnetnie zondirovaniya, v. 2: Moscow, Moscow State Univ. Press.

Koshlakov, G. V., 1970, Influence of a vertical contact on the results of electromagnetic sounding: Geol. i geofiz., v. 11.

Koshlakov, G. V., and Konovalov, Yu. F., 1971, Magnetotelluric sounding in the Tadzhik depression: Prikl. Geofiz., no. 63.

Krasnobaeva, A. G., Vishnev, V. S., and Rudneva, T. L., 1976, Deep electromagnetic studies in the Vrol, *in* Geoelectric and geothermal studies: A. Adam, Ed., Budapest, Akad. Kiado, p. 640-645.

Lipskaya, N. V., Deniskin, N. A., and Nikiforova, N. N., 1976, Deep magnetotelluric surveys in the Belorussian crystalline massif, *in* Geoelectric and geothermal studies: A. Adam, Ed., Budapest, Akad. Kiado, p. 617-619.

Obukhov, G. G., 1976, Possible improvement in the resolution of studies with the magnetotelluric sounding method: Prikl. Geofiz., no. 83.

Podlovilin, E. S., and Gablin, N. S., 1976, Combined magnetotelluric methods and geological problems solved with them in the Timen-Pecher oil and gas province, *in* Elektromagnetnie zondirovaniya, Moscow, Moscow National Univ. Press.

Safonov, A. S., and Bezruk, I. A., 1977, Evaluation of the adequacy of interpreting data from magnetotelluric sounding: Prikl. Geofiz., no. 86, p. 74-79.

Safonov, A. S., Bubnov, V. M., Sysoev, B. K., Chernyavsky, G. A., Chinareva, O. M., and Shaporev, V. A., 1976, Deep magnetotelluric surveys of the Tungus Synclise and on the west Siberian plate, *in* Geoelectric and geothermal studies: A. Adam, Ed., Budapest, Akad. Kiado, p. 666-672.

Shaporev, V. A., and Bubnov, V. M., 1976, Magnetotelluric investigations in the Tungus synclese, *in* Elektromagnetnie zondirovaniya: Moscow, Moscow National Univ. Press.

Sheinkman, A. L., and Lipilin, V. A., 1968, A method of quantitative interpretation of magnetotelluric sounding curves and its use in the geoelectric conditions of the Moscow syneclise: Raxv. Geofiz., no. 30.

Sheriff, R. E., 1973, Encyclopedic dictionary of exploration geophysics: SEG, Tulsa.

Sidorov, V. A., Marinichev, A. P., Bucharskiy, B. V., and Ugarova, N. P., 1976, Basic results of electromagnetic sounding in the Cis-Caspian basin, *in* Elektromagnetnie zondirovaniya, v. 2: Moscow, Moscow National Univ. Press.

Stadnik, Yu. N., Garkusha, I. G., Prokhorov, S. F., Bolashov, E. T., Sen'ko, A. I., 1976, Geological results of magnetotelluric investigations in the White Russian Republic, *in* Electromagnetnie zondirovaniya: Moscow, Moscow State Univ. Press.

Sviridenko, I. L., 1976, Use of the MTS method for studying local structure, *in* Elektromagnetnie zondirovaniya, v. 2: Moscow, Moscow National Univ. Press.

Tomchakov, L. I., and Kharchenko, S. P., 1976, Magnetotelluric sounding with digital recording in the Dneprov-Donetz basin, *in* Electromagnetnie zondirovaniya, v. 2: Moscow, Moscow National Univ. Press.

Tomchakov, L. I., and Kharchenko, S. P., 1976, Use of combined magnetotelluric sounding for studying the morphology of salt structures, *in* Elektromagnetnie zondirovaniya, v. 2: Moscow, Moscow National Univ. Press.

Varlamov, D. A., Konnov, Yu. K., and Feldman, I. S., 1976, Deep magnetotelluric surveys in the eastern Pre-Caucasus, *in* Geoelectric and geothermal studies: A. Adam, Ed., Budapest, Akad. Kiado, p. 646-651.

Vladimirov, N. P., 1976, Deep magnetotelluric surveys in the Baltic-Scandinavian Shield and the Kokchetav Block, *in* Geoelectric and geothermal studies: A. Adam, Ed., Budapest, Akad. Kiado, p. 615-616.

Volarivich, M. P., and Parkhomenko, E. I., 1976, Electrical properties of rocks of high temperatures and pressures, *in* Geoelectric and geothermal studies: A. Adam, Ed., Budapest, Akad. Kiado, p. 321-369.

Yakovlev, I. A., 1964, Magnetotelluric investigations in the Vilyin synclese: Bulletin of Sci. Tech. Inf., no. 1(51), published by Geogeoltekhizdat.

Yakovlev, I. A., Sheinkman, A. L., and Sisoev, B. K., 1975, Quantitative interpretation of magnetotelluric sounding with the use of interval conductance and effective depth of penetration of the electromagnetic field: Prikl. geofiz., no. 70.

Yakovlev, I. A., and Sivitzky, V. P., 1969, A method of quantitative interpretation of MTS curves in the conditions of the geoelectric section of the Viluis syneclise: Rezv. geofiz., no. 36.

——— 1970, On deep structure of the southwestern part of the Viluis syneclise: (from results of the electrical surveys), *in* Tektonika sibiri, v. 3: Nedra Publ. House.